T0275870

Quantum Machine Learning

Quantum Machine Learning

What Quantum Computing Means to Data Mining

Peter Wittek
University of Borås
Sweden

AMSTERDAM • BOSTON • HEIDELBERG • LONDON
NEW YORK • OXFORD • PARIS • SAN DIEGO
SAN FRANCISCO • SINGAPORE • SYDNEY • TOKYO
Academic Press is an imprint of Elsevier

Academic Press is an imprint of Elsevier
525 B Street, Suite 1800, San Diego, CA 92101-4495, USA
225 Wyman Street, Waltham, MA 02451, USA
The Boulevard, Langford Lane, Kidlington, Oxford OX5 1GB, UK
32 Jamestown Road, London NW1 7BY, UK

First edition

British Library Cataloguing-in-Publication Data
A catalogue record for this book is available from the British Library

Library of Congress Cataloging-in-Publication Data
A catalog record for this book is available from the Library of Congress

ISBN: 978-0-12-810040-0

For information on all Elsevier publications
visit our website at store.elsevier.com

Contents

Part Three Quantum Computing and Machine Learning 97

Preface

Machine learning is a fascinating area to work in: from detecting anomalous events in live streams of sensor data to identifying emergent topics involving text collection, exciting problems are never too far away.

Quantum information theory also teems with excitement. By manipulating particles at a subatomic level, we are able to perform Fourier transformation exponentially faster, or search in a database quadratically faster than the classical limit. Superdense coding transmits two classical bits using just one qubit. Quantum encryption is unbreakable—at least in theory.

The fundamental question of this monograph is simple: What can quantum computing contribute to machine learning? We naturally expect a speedup from quantum methods, but what kind of speedup? Quadratic? Or is exponential speedup possible? It is natural to treat any form of reduced computational complexity with suspicion. Are there tradeoffs in reducing the complexity?

Execution time is just one concern of learning algorithms. Can we achieve higher generalization performance by turning to quantum computing? After all, training error is not that difficult to keep in check with classical algorithms either: the real problem is finding algorithms that also perform well on previously unseen instances. Adiabatic quantum optimization is capable of finding the global optimum of nonconvex objective functions. Grover's algorithm finds the global minimum in a discrete search space. Quantum process tomography relies on a double optimization process that resembles active learning and transduction. How do we rephrase learning problems to fit these paradigms?

Storage capacity is also of interest. Quantum associative memories, the quantum variants of Hopfield networks, store exponentially more patterns than their classical counterparts. How do we exploit such capacity efficiently?

These and similar questions motivated the writing of this book. The literature on the subject is expanding, but the target audience of the articles is seldom the academics working on machine learning, not to mention practitioners. Coming from the other direction, quantum information scientists who work in this area do not necessarily aim at a deep understanding of learning theory when devising new algorithms.

This book addresses both of these communities: theorists of quantum computing and quantum information processing who wish to keep up to date with the wider context of their work, and researchers in machine learning who wish to benefit from cutting-edge insights into quantum computing.

I am indebted to Stephanie Wehner for hosting me at the Centre for Quantum Technologies for most of the time while I was writing this book. I also thank Antonio Acín for inviting me to the Institute for Photonic Sciences while I was finalizing the manuscript. I am grateful to Sándor Darányi for proofreading several chapters.

Peter Wittek
Castelldefels, May 30, 2014

Notations

$\mathbb{1}$	indicator function
$\mathbb{C}$	set of complex numbers
d	number of dimensions in the feature space
E	error
$\mathbb{E}$	expectation value
$\mathbf{G}$	group
H	Hamiltonian
$\mathcal{H}$	Hilbert space
I	identity matrix or identity operator
K	number of weak classifiers or clusters, nodes in a neural net
N	number of training instances
P_i	measurement: projective or POVM
$\mathbf{P}$	probability measure
$\mathbb{R}$	set of real numbers
ρ	density matrix
$\sigma_x, \sigma_y, \sigma_z$	Pauli matrices
tr	trace of a matrix
U	unitary time evolution operator
$\mathbf{w}$	weight vector
$\mathbf{x}, \mathbf{x}_i$	data instance
X	matrix of data instances
y, y_i	label
$^\top$	transpose
†	Hermitian conjugate
$\|.\|$	norm of a vector
$[., .]$	commutator of two operators
$\otimes$	tensor product
$\oplus$	XOR operation or direct sum of subspaces

Part One

Fundamental Concepts

Introduction

1

The quest of machine learning is ambitious: the discipline seeks to understand what learning is, and studies how algorithms approximate learning. Quantum machine learning takes these ambitions a step further: quantum computing enrolls the help of nature at a subatomic level to aid the learning process.

Machine learning is based on minimizing a constrained multivariate function, and these algorithms are at the core of data mining and data visualization techniques. The result of the optimization is a decision function that maps input points to output points. While this view on machine learning is simplistic, and exceptions are countless, some form of optimization is always central to learning theory.

The idea of using quantum mechanics for computations stems from simulating such systems. Feynman (1982) noted that simulating quantum systems on classical computers becomes unfeasible as soon as the system size increases, whereas quantum particles would not suffer from similar constraints. Deutsch (1985) generalized the idea. He noted that quantum computers are universal Turing machines, and that quantum parallelism implies that certain probabilistic tasks can be performed faster than by any classical means.

Today, quantum information has three main specializations: quantum computing, quantum information theory, and quantum cryptography (Fuchs, 2002, p. 49). We are not concerned with quantum cryptography, which primarily deals with secure exchange of information. Quantum information theory studies the storage and transmission of information encoded in quantum states; we rely on some concepts such as quantum channels and quantum process tomography. Our primary focus, however, is quantum computing, the field of inquiry that uses quantum phenomena such as superposition, entanglement, and interference to operate on data represented by quantum states.

Algorithms of importance emerged a decade after the first proposals of quantum computing appeared. Shor (1997) introduced a method to factorize integers exponentially faster, and Grover (1996) presented an algorithm to find an element in an unordered data set quadratically faster than the classical limit. One would have expected a slew of new quantum algorithms after these pioneering articles, but the task proved hard (Bacon and van Dam, 2010). Part of the reason is that now we expect that a quantum algorithm should be faster—we see no value in a quantum algorithm with the same computational complexity as a known classical one. Furthermore, even

Quantum Machine Learning. http://dx.doi.org/10.1016/B978-0-12-800953-6.00001-3

with the spectacular speedups, the class NP cannot be solved on a quantum computer in subexponential time (Bennett et al., 1997).

While universal quantum computers remain out of reach, small-scale experiments implementing a few qubits are operational. In addition, quantum computers restricted to domain problems are becoming feasible. For instance, experimental validation of combinatorial optimization on over 500 binary variables on an adiabatic quantum computer showed considerable speedup over optimized classical implementations (McGeoch and Wang, 2013). The result is controversial, however (Rønnow et al., 2014).

Recent advances in quantum information theory indicate that machine learning may benefit from various paradigms of the field. For instance, adiabatic quantum computing finds the minimum of a multivariate function by a controlled physical process using the adiabatic theorem (Farhi et al., 2000). The function is translated to a physical description, the Hamiltonian operator of a quantum system. Then, a system with a simple Hamiltonian is prepared and initialized to the ground state, the lowest energy state a quantum system can occupy. Finally, the simple Hamiltonian is evolved to the target Hamiltonian, and, by the adiabatic theorem, the system remains in the ground state. At the end of the process, the solution is read out from the system, and we obtain the global optimum for the function in question.

While more and more articles that explore the intersection of quantum computing and machine learning are being published, the field is fragmented, as was already noted over a decade ago (Bonner and Freivalds, 2002). This should not come as a surprise: machine learning itself is a diverse and fragmented field of inquiry. We attempt to identify common algorithms and trends, and observe the subtle interplay between faster execution and improved performance in machine learning by quantum computing.

As an example of this interplay, consider convexity: it is often considered a virtue in machine learning. Convex optimization problems do not get stuck in local extrema, they reach a global optimum, and they are not sensitive to initial conditions. Furthermore, convex methods have easy-to-understand analytical characteristics, and theoretical bounds on convergence and other properties are easier to derive. Nonconvex optimization, on the other hand, is a forte of quantum methods. Algorithms on classical hardware use gradient descent or similar iterative methods to arrive at the global optimum. Quantum algorithms approach the optimum through an entirely different, more physical process, and they are not bound by convexity restrictions. Nonconvexity, in turn, has great advantages for learning: sparser models ensure better generalization performance, and nonconvex objective functions are less sensitive to noise and outliers. For this reason, numerous approaches and heuristics exist for nonconvex optimization on classical hardware, which might prove easier and faster to solve by quantum computing.

As in the case of computational complexity, we can establish limits on the performance of quantum learning compared with the classical flavor. Quantum learning is not more powerful than classical learning—at least from an information-theoretic perspective, up to polynomial factors (Servedio and Gortler, 2004). On the other hand, there are apparent computational advantages: certain concept classes are polynomial-time exact-learnable from quantum membership queries, but they are not polynomial-time learnable from classical membership queries (Servedio and Gortler, 2004). Thus quantum machine learning can take logarithmic time in both the number of vectors and their dimension. This is an exponential speedup over classical algorithms, but at the price of having both quantum input and quantum output (Lloyd et al., 2013a).

1.1 Learning Theory and Data Mining

Machine learning revolves around algorithms, model complexity, and computational complexity. Data mining is a field related to machine learning, but its focus is different. The goal is similar: identify patterns in large data sets, but aside from the raw analysis, it encompasses a broader spectrum of data processing steps. Thus, data mining borrows methods from statistics, and algorithms from machine learning, information retrieval, visualization, and distributed computing, but it also relies on concepts familiar from databases and data management. In some contexts, data mining includes any form of large-scale information processing.

In this way, data mining is more applied than machine learning. It is closer to what practitioners would find useful. Data may come from any number of sources: business, science, engineering, sensor networks, medical applications, spatial information, and surveillance, to mention just a few. Making sense of the data deluge is the primary target of data mining.

Data mining is a natural step in the evolution of information systems. Early database systems allowed the storing and querying of data, but analytic functionality was limited. As databases grew, a need for automatic analysis emerged. At the same time, the amount of unstructured information—text, images, video, music—exploded. Data mining is meant to fill the role of analyzing and understanding both structured and unstructured data collections, whether they are in databases or stored in some other form.

Machine learning often takes a restricted view on data: algorithms assume either a geometric perspective, treating data instances as vectors, or a probabilistic one, where data instances are multivariate random variables. Data mining involves preprocessing steps that extract these views from data.

For instance, in text mining—data mining aimed at unstructured text documents—the initial step builds a vector space from documents. This step starts with identification of a set of keywords—that is, words that carry meaning: mainly nouns, verbs, and adjectives. Pronouns, articles, and other connectives are disregarded. Words that occur too frequently are also discarded: these differentiate only a little between two text documents. Then, assigning an arbitrary vector from the canonical basis to each keyword, an indexer constructs document vectors by summing these basis vectors. The summation includes a weighting, where the weighting reflects the relative importance of the keyword in that particular document. Weighting often incorporates the global importance of the keyword across all documents.

The resulting vector space—the term-document space—is readily analyzed by a whole range of machine learning algorithms. For instance, K-means clustering identifies groups of similar documents, support vector machines learn to classify documents to predefined categories, and dimensionality reduction techniques, such as singular value decomposition, improve retrieval performance.

The data mining process often includes how the extracted information is presented to the user. Visualization and human-computer interfaces become important at this stage. Continuing the text mining example, we can map groups of similar documents on a two-dimensional plane with self-organizing maps, giving a visual overview of the clustering structure to the user.

Machine learning is crucial to data mining. Learning algorithms are at the heart of advanced data analytics, but there is much more to successful data mining. While quantum methods might be relevant at other stages of the data mining process, we restrict our attention to core machine learning techniques and their relation to quantum computing.

1.2 Why Quantum Computers?

We all know about the spectacular theoretical results in quantum computing: factoring of integers is exponentially faster and unordered search is quadratically faster than with any known classical algorithm. Yet, apart from the known examples, finding an application for quantum computing is not easy.

Designing a good quantum algorithm is a challenging task. This does not necessarily derive from the difficulty of quantum mechanics. Rather, the problem lies in our expectations: a quantum algorithm must be faster and computationally less complex than any known classical algorithm for the same purpose.

The most recent advances in quantum computing show that machine learning might just be the right field of application. As machine learning usually boils down to a form of multivariate optimization, it translates directly to quantum annealing and adiabatic quantum computing. This form of learning has already demonstrated results on actual quantum hardware, albeit countless obstacles remain to make the method scale further.

We should, however, not confine ourselves to adiabatic quantum computers. In fact, we hardly need general-purpose quantum computers: the task of learning is far more restricted. Hence, other paradigms in quantum information theory and quantum mechanics are promising for learning. Quantum process tomography is able to learn an unknown function within well-defined symmetry and physical constraints—this is useful for regression analysis. Quantum neural networks based on arbitrary implementation of qubits offer a useful level of abstraction. Furthermore, there is great freedom in implementing such networks: optical systems, nuclear magnetic resonance, and quantum dots have been suggested. Quantum hardware dedicated to machine learning may become reality much faster than a general-purpose quantum computer.

1.3 A Heterogeneous Model

It is unlikely that quantum computers will replace classical computers. Why would they? Classical computers work flawlessly at countless tasks, from word processing to controlling complex systems. Quantum computers, on the other hand, are good at certain computational workloads where their classical counterparts are less efficient.

Let us consider the state of the art in high-performance computing. Accelerators have become commonplace, complementing traditional central processing units. These accelerators are good at single-instruction, multiple-data-type parallelism, which is typical in computational linear algebra. Most of these accelerators derive from graphics processing units, which were originally designed to generate three-dimensional images at a high frame rate on a screen; hence, accuracy was not a consideration. With recognition of their potential in scientific computing, the platform evolved to produce high-accuracy double-precision floating point operations. Yet, owing to their design philosophy, they cannot accelerate just any workload. Random data access patterns, for instance, destroy the performance. Inherently single threaded applications will not show competitive speed on such hardware either. In contemporary high-performance computing, we must design algorithms using heterogeneous hardware: some parts execute faster on central processing units, others on accelerators. This model has been so successful that almost all supercomputers being built today include some kind of accelerator.

If quantum computers become feasible, a similar model is likely to follow for at least two reasons:

1. The control systems of the quantum hardware will be classical computers.
2. Data ingestion and measurement readout will rely on classical hardware.

More extensive collaboration between the quantum and classical realms is also expected. Quantum neural networks already hint at a recursive embedding of classical and quantum computing (Section 11.3). This model is the closest to the prevailing standards of high-performance computing: we already design algorithms with accelerators in mind.

1.4 An Overview of Quantum Machine Learning Algorithms

Dozens of articles have been published on quantum machine learning, and we observe some general characteristics that describe the various approaches. We summarize our observations in Table 1.1, and detail the main traits below.

Many quantum learning algorithms rely on the application of Grover's search or one of its variants (Section 4.5). This includes mostly unsupervised methods: K-medians, hierarchical clustering, or quantum manifold embedding (Chapter 10). In addition, quantum associative memory and quantum neural networks often rely on this search (Chapter 11). An early version of quantum support vector machines also

Table 1.1 The Characteristics of the Main Approaches to Quantum Machine Learning

Algorithm	Reference	Grover	Speedup	Quantum Data	Generalization Performance	Implementation
K-medians	Aïmeur et al. (2013)	Yes	Quadratic	No	No	No
Hierarchical clustering	Aïmeur et al. (2013)	Yes	Quadratic	No	No	No
K-means	Lloyd et al. (2013a)	Optional	Exponential	Yes	No	No
Principal components	Lloyd et al. (2013b)	No	Exponential	Yes	No	No
Associative memory	Ventura and Martinez (2000)	Yes		No	No	No
	Trugenberger (2001)	No		No	No	No
Neural networks	Narayanan and Menneer (2000)	Yes		No	Numerical	Yes
Support vector machines	Anguita et al. (2003)	Yes	Quadratic	No	Analytical	No
	Rebentrost et al. (2013)	No	Exponential	Yes	No	No
Nearest neighbors	Wiebe et al. (2014)	Yes	Quadratic	No	Numerical	No
Regression	Bisio et al. (2010)	No		Yes	No	No
Boosting	Neven et al. (2009)	No	Quadratic	No	Analytical	Yes

The column headed "Algorithm" lists the classical learning method. The column headed "Reference" lists the most important articles related to the quantum variant. The column headed "Grover" indicates whether the algorithm uses Grover's search or an extension thereof. The column headed "Speedup" indicates how much faster the quantum variant is compared with the best known classical version. "Quantum data" refers to whether the input, output, or both are quantum states, as opposed to states prepared from classical vectors. The column headed "Generalization performance" states whether this quality of the learning algorithm was studied in the relevant articles. "Implementation" refers to attempts to develop a physical realization.

uses Grover's search (Section 12.2). In total, about half of all the methods proposed for learning in a quantum setting use this algorithm.

Grover's search has a quadratic speedup over the best possible classical algorithm on unordered data sets. This sets the limit to how much faster those learning methods that rely on it get. Exponential speedup is possible in scenarios where both the input and the output are also quantum: listing class membership or reading the classical data once would imply at least linear time complexity, which could only be a polynomial speedup. Examples include quantum principal component analysis (Section 10.3), quantum K-means (Section 10.5), and a different flavor of quantum support vector machines (Section 12.3). Regression based on quantum process tomography requires an optimal input state, and, in this regard, it needs a quantum input (Chapter 13). At a high level, it is possible to define an abstract class of problems that can only be learned in polynomial time by quantum algorithms using quantum input (Section 2.5).

A strange phenomenon is that few authors have been interested in the generalization performance of quantum learning algorithms. Analytical investigations are especially sparse, with quantum boosting by adiabatic quantum computing being a notable exception (Chapter 14), along with a form of quantum support vector machines (Section 12.2). Numerical comparisons favor quantum methods in the case of quantum neural networks (Chapter 11) and quantum nearest neighbors (Section 12.1).

While we are far from developing scalable universal quantum computers, learning methods require far more specialized hardware, which is more attainable with current technology. A controversial example is adiabatic quantum optimization in learning problems (Section 14.7), whereas more gradual and well founded are small-scale implementations of quantum perceptrons and neural networks (Section 11.4).

1.5 Quantum-Like Learning on Classical Computers

Machine learning has a lot to adopt from quantum mechanics, and this statement is not restricted to actual quantum computing implementations of learning algorithms. Applying principles from quantum mechanics to design algorithms for classical computers is also a successful field of inquiry. We refer to these methods as quantum-like learning. Superposition, sensitivity to contexts, entanglement, and the linearity of evolution prove to be useful metaphors in many scenarios. These methods are outside our scope, but we highlight some developments in this section. For a more detailed overview, we refer the reader to Manju and Nigam (2012).

Computational intelligence is a field related to machine learning that solves optimization problems by nature-inspired computational methods. These include swarm intelligence (Kennedy and Eberhart, 1995), force-driven methods (Chatterjee et al., 2008), evolutionary computing (Goldberg, 1989), and neural networks (Rumelhart et al., 1994). A new research direction which borrows metaphors from quantum physics emerged over the past decade. These quantum-like methods in machine learning are in a way inspired by nature; hence, they are related to computational intelligence.

Quantum-like methods have found useful applications in areas where the system is displaying contextual behavior. In such cases, a quantum approach naturally incorporates this behavior (Khrennikov, 2010; Kitto, 2008). Apart from contextuality, entanglement is successfully exploited where traditional models of correlation fail (Bruza and Cole, 2005), and quantum superposition accounts for unusual results of combining attributes of data instances (Aerts and Czachor, 2004).

Quantum-like learning methods do not represent a coherent whole; the algorithms are liberal in borrowing ideas from quantum physics and ignoring others, and hence there is seldom a connection between two quantum-like learning algorithms.

Coming from evolutionary computing, there is a quantum version of particle swarm optimization (Sun et al., 2004). The particles in a swarm are agents with simple patterns of movements and actions, each one is associated with a potential solution. Relying on only local information, the quantum variant is able to find the global optimum for the optimization problem in question.

Dynamic quantum clustering emerged as a direct physical metaphor of evolving quantum particles (Weinstein and Horn, 2009). This approach approximates the potential energy of the Hamiltonian, and evolves the system iteratively to identify the clusters. The great advantage of this method is that the steps can be computed with simple linear algebra operations. The resulting evolving cluster structure is similar to that obtained with a flocking-based approach, which was inspired by biological systems (Cui et al., 2006), and it is similar to that resulting from Newtonian clustering with its pairwise forces (Blekas and Lagaris, 2007). Quantum-clustering-based support vector regression extends the method further (Yu et al., 2010).

Quantum neural networks exploit the superposition of quantum states to accommodate gradual membership of data instances (Purushothaman and Karayiannis, 1997). Simulated quantum annealing avoids getting trapped in local minima by using the metaphor of quantum tunneling (Sato et al., 2009)

The works cited above highlight how the machine learning community may benefit from quantum metaphors, potentially gaining higher accuracy and effectiveness. We believe there is much more to gain. An attractive aspect of quantum theory is the inherent structure which unites geometry and probability theory in one framework. Reasoning and learning in a quantum-like method are described by linear algebra operations. This, in turn, translates to computational advantages: software libraries of linear algebra routines are always the first to be optimized for emergent hardware. Contemporary high-performance computing clusters are often equipped with graphics processing units, which are known to accelerate many computations, including linear algebra routines, often by several orders of magnitude. As pointed out by Asanovic et al. (2006), the overarching goal of the future of high-performance computing should be to make it easy to write programs that execute efficiently on highly parallel computing systems. The metaphors offered by quantum-like methods bring exactly this ease of programming supercomputers to machine learning. Early results show that quantum-like methods can, indeed, be accelerated by several orders of magnitude (Wittek, 2013).

Machine Learning

2

Machine learning is a field of artificial intelligence that seeks patterns in empirical data without forcing models on the data—that is, the approach is data-driven, rather than model-driven (Section 2.1). A typical example is clustering: given a distance function between data instances, the task is to group similar items together using an iterative algorithm. Another example is fitting a multidimensional function on a set of data points to estimate the generating distribution.

Rather than a well-defined field, machine learning refers to a broad range of algorithms. A feature space, a mathematical representation of the data instances under study, is at the heart of learning algorithms. Learning patterns in the feature space may proceed on the basis of statistical models or other methods known as algorithmic learning theory (Section 2.2).

Statistical modeling makes propositions about populations, using data drawn from the population of interest, relying on a form of random sampling. Any form of statistical modeling requires some assumptions: a statistical model is a set of assumptions concerning the generation of the observed data and similar data (Cox, 2006).

This contrasts with methods from algorithmic learning theory, which are not statistical or probabilistic in nature. The advantage of algorithmic learning theory is that it does not make use of statistical assumptions. Hence, we have more freedom in analyzing complex real-life data sets, where samples are dependent, where there is excess noise, and where the distribution is entirely unknown or skewed.

Irrespective of the approach taken, machine learning algorithms fall into two major categories (Section 2.3):

1. Supervised learning: the learning algorithm uses samples that are labeled. For example, the samples are microarray data from cells, and the labels indicate whether the sample cells are cancerous or healthy. The algorithm takes these labeled samples and uses them to induce a classifier. This classifier is a function that assigns labels to samples, including those that have never previously been seen by the algorithm.
2. Unsupervised learning: in this scenario, the task is to find structure in the samples. For instance, finding clusters of similar instances in a growing collection of text documents reveals topical changes across time, highlighting trends of discussions, and indicating themes that are dropping out of fashion.

Learning algorithms, supervised or unsupervised, statistical or not statistical, are expected to generalize well. Generalization means that the learned structure will apply

Quantum Machine Learning. http://dx.doi.org/10.1016/B978-0-12-800953-6.00002-5

beyond the training set: new, unseen instances will get the correct label in supervised learning, or they will be matched to their most likely group in unsupervised learning. Generalization usually manifests itself in the form of a penalty for complexity, such as restrictions for smoothness or bounds on the vector space norm. Less complex models are less likely to overfit the data (Sections 2.4 and 2.5).

There is, however, no free lunch: without a priori knowledge, finding a learning model in reasonable computational time that applies to all problems equally well is unlikely. For this reason, the combination of several learners is commonplace (Section 2.6), and it is worth considering the computational complexity in learning theory (Section 2.7).

While there are countless other important issues in machine learning, we restrict our attention to the ones outlined in this chapter, as we deem them to be most relevant to quantum learning models.

2.1 Data-Driven Models

Machine learning is an interdisciplinary field: it draws on traditional artificial intelligence and statistics. Yet, it is distinct from both of them.

Statistics and statistical inference put data at the center of analysis to draw conclusions. Parametric models of statistical inference have strong assumptions. For instance, the distribution of the process that generates the observed values is assumed to be a multivariate normal distribution with only a finite number of unknown parameters. Nonparametric models do not have such an assumption. Since incorrect assumptions invalidate statistical inference (Kruskal, 1988), nonparametric methods are always preferred. This approach is closer to machine learning: fewer assumptions make a learning algorithm more general and more applicable to multiple types of data.

Deduction and reasoning are at the heart of artificial intelligence, especially in the case of symbolic approaches. Knowledge representation and logic are key tools. Traditional artificial intelligence is thus heavily dependent on the model. Dealing with uncertainty calls for statistical methods, but the rigid models stay. Machine learning, on the other hand, allows patterns to emerge from the data, whereas models are secondary.

2.2 Feature Space

We want a learning algorithm to reveal insights into the phenomena being observed. A feature is a measurable heuristic property of the phenomena. In the statistical literature, features are usually called independent variables, and sometimes they are referred to as explanatory variables or predictors. Learning algorithms work with features—a careful selection of features will lead to a better model.

Features are typically numeric. Qualitative features—for instance, string values such as small, medium, or large—are mapped to numeric values. Some discrete structures, such as graphs (Kondor and Lafferty, 2002) or strings (Lodhi et al., 2002), have nonnumeric features.

Good features are discriminating: they aid the learner in identifying patterns and distinguishing between data instances. Most algorithms also assume independent features with no correlation between them. In some cases, dependency between features is beneficial, especially if only a few features are nonzero for each data instance—that is, the features are sparse (Wittek and Tan, 2011).

The multidisciplinary nature of machine learning is reflected in how features are viewed. We may take a geometric view, treating features as tuples, vectors in a high-dimensional space—the feature space. Alternatively, we may view features from a probabilistic perspective, treating them as a multivariate random variables.

In the geometric view, features are grouped into a feature vector. Let d denote the number of features. One vector of the canonical basis $\{\mathbf{e}_1, \mathbf{e}_2, \ldots, \mathbf{e}_d\}$ of $\mathbb{R}^d$ is assigned to each feature. Let x_{ij} be the weight of a feature i in data instance j. Thus, a feature vector $\mathbf{x}_j$ for the object j is a linear combination of the canonical basis vectors:

$$\mathbf{x}_j = \sum_{i=1}^{d} x_{ij}\mathbf{e}_i. \tag{2.1}$$

By writing $\mathbf{x}_j$ as a column vector, we have $\mathbf{x}_j^\top = (x_{1j}, x_{2j}, \ldots, x_{dj})$. For a collection of N data instances, the x_{ij} weights form a $d \times N$ matrix.

Since the basis vectors of the canonical basis are perpendicular to one another, this implies the assumption that the features are mutually independent; this assumption is often violated. The assignment of features to vectors is arbitrary: a feature may be assigned to any of the vectors of the canonical basis.

With use of the geometric view, distance functions, norms of vectors, and angles help in the design of learning algorithms. For instance, the Euclidean distance is commonly used, and it is defined as follows:

$$d(\mathbf{x}_i, \mathbf{x}_j) = \sqrt{\sum_{k=1}^{d} (\mathbf{x}_{ki} - \mathbf{x}_{kj})^2}. \tag{2.2}$$

If the feature space is binary, we often use the Hamming distance, which measures how many 1's are different in the two vectors:

$$d(\mathbf{x}_i, \mathbf{x}_j) = \sum_{k=1}^{d} (\mathbf{x}_{ki} \oplus \mathbf{x}_{kj}), \tag{2.3}$$

where $\oplus$ is the XOR operator. This distance is useful in efficiently retrieving elements from a quantum associative memory (Section 11.1).

The cosine of the smallest angle between two vectors, also called the cosine similarity, is given as

$$\cos(\mathbf{x}_i, \mathbf{x}_j) = \frac{\mathbf{x}_i^\top \mathbf{x}_j}{\|\mathbf{x}_i\| \, \|\mathbf{x}_j\|}. \tag{2.4}$$

Other distance and similarity functions are of special importance in kernel-based learning methods (Chapter 7).

The probabilistic view introduces a different set of tools to help design algorithms. It assumes that each feature is a random variable, defined as a function that assigns a real number to every outcome of an experiment (Zaki and Meira, 2013, p. 17). A discrete random variable takes any of a specified finite or countable list of values. The associated probabilities form a probability mass function. A continuous random variable takes any numerical value in an interval or in a collection of intervals. In the continuous case, a probability density function describes the distribution.

Irrespective of the type of random variable, the associated cumulative probabilities must add up to 1. In the geometric view, this corresponds to normalization constraints.

Like features group into a feature vector in the geometric view, the probabilistic view has a multivariate random variable for each data instance: $(X_1, X_2, \ldots, X_d)^\top$. A joint probability mass function or density function describes the distribution. The random variables are independent if and only if the joint probability decomposes to the product of the constituent distributions for every value of the range of the random variables:

$$\mathbf{P}(X_1, X_2, \ldots, X_d) = \mathbf{P}(X_1)\mathbf{P}(X_2) \cdots \mathbf{P}(X_d). \tag{2.5}$$

This independence translates to the orthogonality of the basis vectors in the geometric view.

Not all features are equally important in the feature space. Some might mirror the distribution of another one—strong correlations may exist among features, violating independence assumptions. Others may get consistently low weights or low probabilities to the extent that their presence is negligible. Having more features should result in more discriminating power and thus higher effectiveness in machine learning. However, practical experience with machine learning algorithms has shown that this is not always the case.

Irrelevant or redundant training information adversely affects many common machine learning algorithms. For instance, the nearest neighbor algorithm is sensitive to irrelevant features. Its sample complexity—number of training examples needed to reach a given accuracy level—grows exponentially with the number of irrelevant features (Langley and Sage, 1994b). Sample complexity for decision tree algorithms grows exponentially for some concepts as well. Removing irrelevant and redundant information produces smaller decision trees (Kohavi and John, 1997). The naïve Bayes classifier is also affected by redundant features owing to its assumption that features are independent given the class label (Langley and Sage, 1994a). However, in the case of support vector machines, feature selection has a smaller impact on the efficiency (Weston et al., 2000).

The removal of redundant features reduces the number of dimensions in the space, and may improve generalization performance (Section 2.4). The potential benefits of feature selection and feature extraction include facilitating data visualization and data understanding, reducing the measurement and storage requirements, reducing training and utilization times, and defying the curse of dimensionality to improve prediction performance (Guyon et al., 2003). Methods differ in which aspect they put more emphasis on. Getting the right number of features is a hard task.

Feature selection and feature extraction are the two fundamental approaches in reducing the number of dimensions. Feature selection is the process of identifying and removing as much irrelevant and redundant information as possible. Feature extraction, on the other hand, creates a new, reduced set of features which combines elements of the original feature set.

A feature selection algorithm employs an evaluation measure to score different subsets of the features. For instance, feature wrappers take a learning algorithm, and train it on the data using subsets of the feature space. The error rate will serve as an evaluation measure. Since feature wrappers train a model in every step, they are expensive to evaluate. Feature filters use more direct evaluation measures such as correlation or mutual information. Feature weighting is a subclass of feature filters. It does not reduce the actual dimension, but weights and ranks features according to their importance.

Feature extraction applies a transformation on the feature vector to perform dimensionality reduction. It often takes the form of a projection: principal component analysis and lower-rank approximation with singular value decomposition belong to this category. Nonlinear embeddings are also popular. The original feature set will not be present, and only derived features that are optimal according to some measure will be present—this task may be treated as an unsupervised learning scenario (Section 2.3).

2.3 Supervised and Unsupervised Learning

We often have a well-defined goal for learning. For instance, taking a time series, we want a learning algorithm to fit a nonlinear function to approximate the generating process. In other cases, the objective of learning is less obvious: there is a pattern we are seeking, but we are uncertain what it might be. Given a set of high-dimensional points, we may ask which points form nonoverlapping groups—clusters. The clusters and their labels are unknown before we begin. According to whether the goal is explicit, machine learning splits into two major paradigms: supervised and unsupervised learning.

In supervised learning, each data point in a feature space comes with a label (Figure 2.1). The label is also called an output or a response, or, in classical statistical literature, a dependent variable. Labels may have a continuous numerical range, leading to a regression problem. In classification, the labels are the elements of a fixed, finite set of numerical values or qualitative descriptors. If the set has two values—for instance, yes or no, 0 or 1, $+1$ or -1—we call the problem binary classification. Multiclass problems have more than two labels. Qualitative labels are typically encoded as integers.

A supervised learner predicts the label of instances after training on a sample of labeled examples, the training set. At a high level, supervised learning is about fitting a

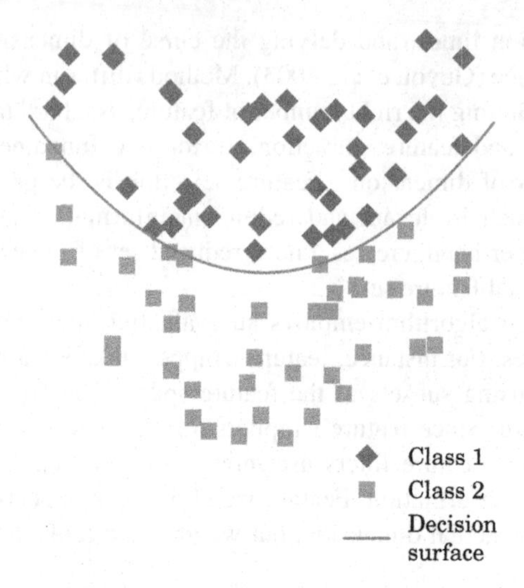

Figure 2.1 Supervised learning. Given labeled training instances, the goal is to identify a decision surface that separates the classes.

predefined multivariate function to a set of points. In other words, supervised learning is function approximation.

We denote a label by y. The training set is thus a collection of pairs of data points and corresponding labels: $\{(\mathbf{x}_1, y_1), (\mathbf{x}_2, y_2), \ldots, (\mathbf{x}_N, y_N)\}$, where N is the number of training instances.

In an unsupervised scenario, the labels are missing. A learning algorithm must extract structure in the data on its own (Figure 2.2). Clustering and low-dimensional embedding belong to this category. Clustering finds groups of data instances such that instances in the same group are more similar to each other than to those in other groups. The groups—or clusters—may be embedded in one another, and the density of data instances often varies across the feature space; thus, clustering is a hard problem to solve in general.

Low-dimensional embedding involves projecting data instances from the high-dimensional feature space to a more manageable number of dimensions. The target number of dimensions depends on the task. It can be as high as 200 or 300. For example, if the feature space is sparse, but it has several million dimensions, it is advantageous to embed the points in 200 dimensions (Deerwester et al., 1990). If we project to just two or three dimensions, we can plot the data instances in the embedding space to reveal their topology. For this reason, a good embedding algorithm will preserve either the local topology or the global topology of the points in the original high-dimensional space.

Semisupervised learning makes use of both labeled and unlabeled examples to build a model. Labels are often expensive to obtain, whereas data instances are available in abundance. The semisupervised approach learns the pattern using the labeled examples, then refines the decision boundary between the classes with the unlabeled examples.

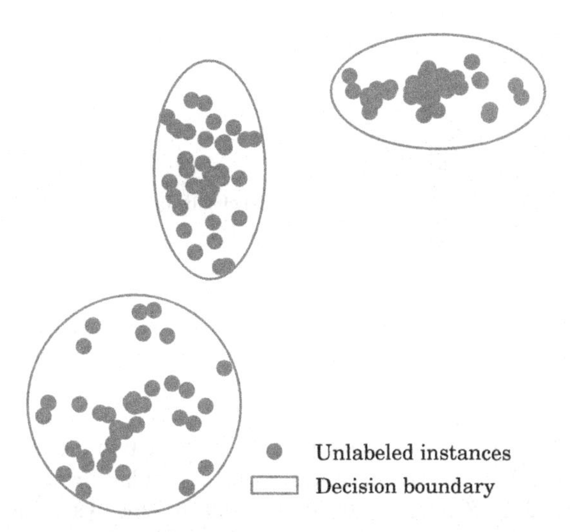

Figure 2.2 Unsupervised learning. The training instances do not have a label. The learning process identifies the classes automatically, often creating a decision boundary.

Active learning is a variant of semisupervised learning in which the learning algorithm is able to solicit labels for problematic unlabeled instances from an appropriate information source—for instance, from a human annotator (Settles, 2009). Similarly to the semisupervised setting, there are some labels available, but most of the examples are unlabeled. The task in a learning iteration is to choose the optimal set of unlabeled examples for which the algorithm solicits labels. Following Settles (2009), these are some typical strategies to identify the set for labeling:

- Uncertainty sampling: the selected set corresponds to those data instances where the confidence is low.
- Query by committee: train a simple ensemble (Section 2.6) that casts votes on data instances, and select those which are most ambiguous.
- Expected model change: select those data instances that would change the current model the most if the learner knew its label. This approach is particularly fruitful in gradient-descent-based models, where the expected change is easy to quantify by the length of the gradient.
- Expected error reduction: select those data instances where the model performs poorly—that is, where the generalization error (Section 2.4) is most likely to be reduced.
- Variance reduction: generalization performance is hard to measure, whereas minimizing output variance is far more feasible; select those data instances which minimize output variance.
- Density-weighted methods: the selected instances should be not only uncertain, but also representative of the underlying distribution.

It is interesting to contrast these active learning strategies with the selection of optimal state in quantum process tomography (Section 13.6).

One particular form of learning, transductive learning, will be relevant in later chapters, most notably in Chapter 13. The models mentioned so far are inductive: on the basis of data points—labeled or unlabeled—we infer a function

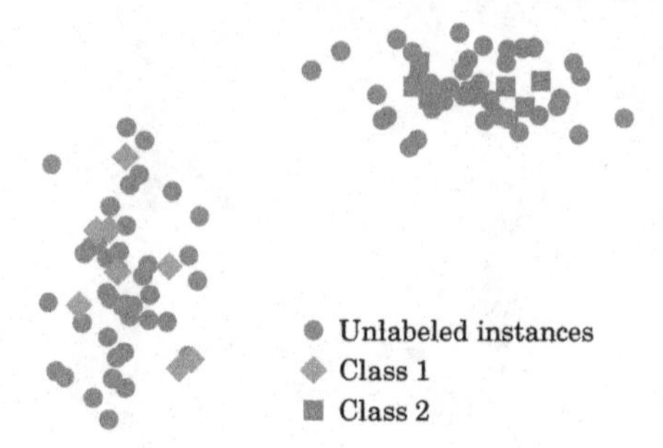

Figure 2.3 Transductive learning. A model is not inferred, there are no decision surfaces. The label of training instances is propagated to the unlabeled instances, which are provided at the same time as the training instances.

that will be applied to unseen data points. Transduction avoids this inference to the more general case, and it infers from particular instances to particular instances (Figure 2.3) (Gammerman et al., 1998). This way, transduction asks for less: an inductive function implies a transductive one. Transduction is similar to instance-based learning, a family of algorithms that compares new problem instances with training instances—K-means clustering is an example (Section 5.3). If some labels are available, transductive learning is similar to semisupervised learning. Yet, transduction is different from all the learning approaches mentioned thus far. Instance-based learning can be inductive, and semisupervised learning is inductive, whereas transductive learning avoids inductive reasoning by definition.

2.4 Generalization Performance

If a learning algorithm learns to reproduce the labels of the training data with 100% accuracy, it still does not follow that the learned model will be useful. What makes a good learner? A good algorithm will generalize well to previously unseen instances. This is why we start training an algorithm: it is hardly interesting to see labeled examples classified again. Generalization performance characterizes a learner's prediction capability on independent test data.

Consider a family of functions f that approximate a function that generates the data $g(\mathbf{x}) = y$ based on a sample $\{(\mathbf{x}_1, y_1), (\mathbf{x}_2, y_2), \ldots, (\mathbf{x}_N, y_N)\}$. The sample itself suffers from random noise with a zero mean and variance σ^2.

We define a loss function L depending on the values y takes. If y is a continuous real number—that is, we have a regression problem—typical choices are the squared error

$$L(y_i, f(\mathbf{x}_i)) = (y_i - f(\mathbf{x}_i))^2, \tag{2.6}$$

and the absolute error

$$L(y_i, f(\mathbf{x}_i)) = |y_i - f(\mathbf{x}_i)|. \tag{2.7}$$

In the case of binary classes, the 0-1 loss function is defined as

$$L(y_i, f(\mathbf{x}_i)) = \mathbb{1}_{\{f(\mathbf{x}_i) \neq y_i\}}, \tag{2.8}$$

where $\mathbb{1}$ is the indicator function. Optimizing for a classification problem with a 0-1 loss function is an NP-hard problem even for such a relatively simple class of functions as linear classifiers (Feldman et al., 2012). It is often approximated by a convex function that makes optimization easier. The hinge loss—notable for its use by support vector machines—is one such approximation:

$$L(y_i, f(\mathbf{x}_i)) = \max(0, 1 - f(\mathbf{x}_i)y). \tag{2.9}$$

Here $f : \mathbb{R}^d \mapsto \mathbb{R}$—that is, the range of the function is not just $\{0, 1\}$.

Given a loss function, the training error (or empirical risk) is defined as

$$E = \frac{1}{n} \sum_{i=1}^{n} L(y_i, f(\mathbf{x}_i)). \tag{2.10}$$

Finding a model in a class of functions that minimizes this error is called empirical risk minimization. A model with zero training error, however, overfits the training data and will generalize poorly. Consider, for instance, the following function:

$$f(\mathbf{x}) = \begin{cases} y_i & \text{if } \mathbf{x} = \mathbf{x}_i, \\ 0 & \text{otherwise.} \end{cases} \tag{2.11}$$

This function is empirically optimal—the training error is zero. Yet, it is easy to see that this function is not what we are looking for.

Take a test sample $\mathbf{x}$ from the underlying distribution. Given the training set, the test error or generalization error is

$$E_{\mathbf{x}}(f) = L(\mathbf{x}, f(\mathbf{x})). \tag{2.12}$$

The expectation value of the generalization error is the true error we are interested in:

$$E_N(f) = \mathbb{E}(L(\mathbf{x}, f(\mathbf{x})) | \{(\mathbf{x}_1, y_1), (\mathbf{x}_2, y_2), \ldots, (\mathbf{x}_N, y_N)\}). \tag{2.13}$$

We estimate the true error over test samples from the underlying distribution.

Let us analyze the structure of the error further. The error over the distribution will be $E^* = \mathbb{E}[L(\mathbf{x}, f(\mathbf{x}))] = \sigma^2$; this error is also called Bayes error. The best possible model of the family of functions f will have an error that no longer depends on the training set: $E_{\text{best}}(f) = \inf\{\mathbb{E}[L(\mathbf{x}, f(\mathbf{x}))]\}$.

The ultimate question is how close we can get with the family of functions to the Bayes error using the sample:

$$E_N(f) - E^* = (E_N(f) - E_{\text{best}}(f)) + (E_{\text{best}}(f) - E^*). \tag{2.14}$$

The first part of the sum is the estimation error: $E_N(f) - E_{\text{best}}(f)$. This is controlled and usually small.

The second part is the approximation error or model bias: $E_{\text{best}}(f) - E^*$. This is characteristic for the family of approximating functions chosen, and it is harder to control, and typically larger than the estimation error.

The estimation error and model bias are intrinsically linked. The more complex we make the model f, the lower the bias is, but in exchange, the estimation error increases. This tradeoff is analyzed in Section 2.5.

2.5 Model Complexity

The complexity of the class of functions performing classification or regression and the algorithm's generalizability are related. The Vapnik-Chervonenkis (VC) theory provides a general measure of complexity and proves bounds on errors as a function of complexity. Structural risk minimization is the minimization of these bounds, which depend on the empirical risk and the capacity of the function class (Vapnik, 1995).

Consider a function f with a parameter vector θ: it shatters a set of data points $\{\mathbf{x}_1, \mathbf{x}_2, \ldots, \mathbf{x}_N\}$ if, for all assignments of labels to those points, there exists a θ such that the function f makes no errors when evaluating that set of data points. A set of N points can be labeled in 2^N ways. A rich function class is able to realize all 2^N separations—that is, it shatters the N points.

The idea of VC dimensions lies at the core of the structural risk minimization theory: it measures the complexity of a class of functions. This is in stark contrast to the measures of generalization performance in Section 2.4, which derive them from the sample and the distribution.

The VC dimension of a function f is the maximum number of points that are shattered by f. In other words, the VC dimension of the function f is h', where h' is the maximum h such that some data point set of cardinality h can be shattered by f. The VC dimension can be infinity (Figure 2.4).

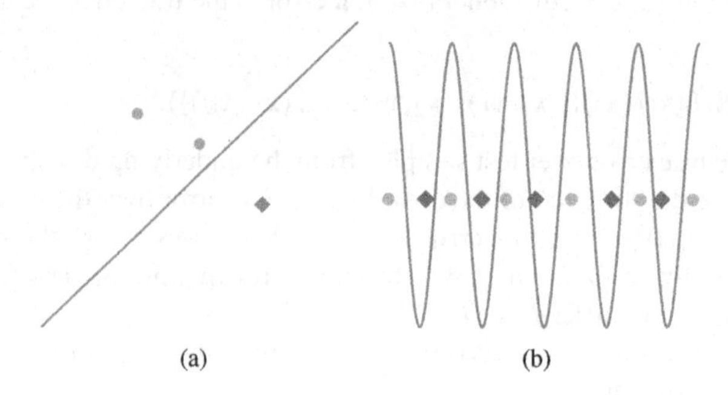

Figure 2.4 Examples of shattering sets of points. (a) A line on a plane can shatter a set of three points with arbitrary labels, but it cannot shatter certain sets of four points; hence, a line has a VC dimension of four. (b) A sine function can shatter any number of points with any assignment of labels; hence, its VC dimension is infinite.

Vapnik's theorem proves a connection between the VC dimension, empirical risk, and the generalization performance (Vapnik and Chervonenkis, 1971). The probability of the test error distancing from an upper bound on data that are drawn independent and identically distributed from the same distribution as the training set is given by

$$P\left(E_N(f) \leq E + \sqrt{\frac{h[\log(2n/h)+1] - \log(\eta/4)}{n}}\right) = 1 - \eta \tag{2.15}$$

if $h \ll n$, where h is the VC dimension of the function. When $h \ll n$, the function class should be large enough to provide functions that are able to model the hidden dependencies in the joint distribution $\mathbf{P}(\mathbf{x}, y)$.

This theorem formally binds model complexity and generalization performance. Empirical risk minimization—introduced in Section 2.4—allows us to pick an optimal model given a fixed VC dimension h for the function class. The principle that derives from Vapnik's theorem—structural risk minimization—goes further. We optimize empirical risk for a nested sequence of increasingly complex models with VC dimensions $h_1 < h_2 < \cdots$, and select the model with the smallest value of the upper bound in Equation 2.15.

The VC dimension is a one-number summary of the learning capacity of a class of functions, which may prove crude for certain classes (Schölkopf and Smola, 2001, p. 9). Moreover, the VC dimension is often difficult to calculate. Structural risk minimization successfully applies in some cases, such as in support vector machines (Chapter 7).

A concept related to VC dimension is probably approximately correct (PAC) learning (Valiant, 1984). PAC learning stems from a different background: it introduces computational complexity to learning theory. Yet, the core principle is common. Given a finite sample, a learner has to choose a function from a given class such that, with high probability, the selected function will have low generalization error. A set of labels y_i are PAC-learnable if there is an algorithm that can approximate the labels with a predefined error $0 < \epsilon < 1/2$ with a probability at least $1 - \delta$, where $0 < \delta < 1/2$ is also predefined. A problem is efficiently PAC-learnable if it is PAC-learnable by an algorithm that runs in time polynomial in $1/\epsilon$, $1/\delta$, and the dimension d of the instances. Under some regularity conditions, a problem is PAC-learnable if and only if its VC dimension is finite (Blumer et al., 1989).

An early result in quantum learning theory proved that all PAC-learnable function classes are learnable by a quantum model (Servedio and Gortler, 2001); in this sense, quantum and classical PAC learning are equivalent. The lower bound on the number of examples required for quantum PAC learning is close to the classical bound (Atici and Servedio, 2005). Certain classes of functions with noisy labels that are classically not PAC-learnable can be learned by a quantum model (Bshouty and Jackson, 1995). If we restrict our attention to transductive learning problems, and we do not want to generalize to a function that would apply to an arbitrary number of new instances, we can explicitly define a class of problems that would take an exponential amount of time to solve classically, but a quantum algorithm could learn it in polynomial time (Gavinsky, 2012). This approach does not fall in the bounded error quantum polynomial time class of decision problems, to which most known quantum algorithms belong (see Section 4.6).

The connection between PAC-learning theory and machine learning is indirect, but explicit connection has been made to some learning algorithms, including neural networks (Haussler, 1992). This already suggests that quantum machine learning algorithms learn with a higher precision, even in the presence of noise. We give more specific details in Chapters 11 and 14. Here we point out that we do not deal with the exact identification of a function (Angluin, 1988), which also has various quantum formulations and accompanying literature.

Irrespective of how we optimize the learning function, there is no free lunch: there cannot be a class of functions that is optimal for all learning problems (Wolpert and Macready, 1997). For any optimization or search algorithm, better performance in one class of problems is balanced by poorer performance in another class. For this reason alone, it is worth looking into combining different learning models.

2.6 Ensembles

A learning algorithm will always have strengths and weaknesses: a single model is unlikely to fit every possible scenario. Ensembles combine multiple models to achieve higher generalization performance than any of the constituent models is capable of. A constituent model is also called a base classifier or weak learner, and the composite model is called a strong learner.

Apart from generalization performance, there are further reasons for using ensemble-based systems (Polikar, 2006):

- Large volumes of data: the computational complexity of many learning algorithms is much higher than linear time. Large data sets are often not feasible for training an algorithm. Splitting the data, training separate classifiers, and using an ensemble of them is often more efficient.
- Small volumes of data: ensembles help with the other extreme as well. By resampling with replacement, numerous classifiers learn on samples of the same data, yielding a higher performance.
- Divide and conquer: the decision boundary of problems is often a complex nonlinear surface. Instead of using an intricate algorithm to approximate the boundary, several simple learners might work just as efficiently.
- Data fusion: data often originate from a range of sources, leading to vastly different feature sets. Some learning algorithms work better with one type of feature set. Training separate algorithms on divisions of feature sets leads to data fusion, and efficient composite learners.

Ensembles yield better results when there is considerable diversity among the base classifiers—irrespective of the measure of diversity (Kuncheva and Whitaker, 2003). If diversity is sufficient, base classifiers make different errors, and a strategic combination may reduce the total error—ideally improving generalization performance.

The generic procedure of ensemble methods has two steps: first, develop a set of base classifiers from the training data; second, combine them to form a composite predictor. In a simple combination, the base learners vote, and the label prediction is based on the collection of votes. More involved methods weigh the votes of the base learners.

More formally, we train K base classifiers, $M_1, M_2, \ldots, M_K$. Each model is trained on a subset of $\{(\mathbf{x}_1, y_1), (\mathbf{x}_2, y_2), \ldots, (\mathbf{x}_N, y_N)\}$; the subsets may overlap in consecutive training runs. A base classifier should have higher accuracy than random guessing. The training of an M_i classifier is independent from training of the other classifiers; hence, parallelization is easy and efficient (Han et al., 2012, p. 378).

Popular ensemble methods include bagging, random forests, stacking, and boosting. In bagging—short for "bootstrap aggregating"—the base learners vote with equal weight (Breiman, 1996; Efron, 1979). To improve diversity among the learned models, bagging generates a random training subset from the data for each base classifier M_i.

Random forests are an application of bagging to decision trees (Breiman, 2001). Decision trees are simple base classifiers that are fast to train. Random forests train many decision trees on random samples of the data, keeping the complexity of each tree low. Bagging decides the eventual label on a data instance. Random forests are known to be robust to noise.

Stacking is an improvement over bagging. Instead of counting votes, stacking trains a learner on the basis of the output of the base classifiers (Wolpert, 1992). For instance, suppose that the decision surface of a particular base classifier cannot fit a part of the data and it incorrectly learns a certain region of the feature space. Instances coming from that region will be consistently misclassified: the stacked learner may be able to learn this pattern, and correct the result.

Unlike the previous methods, boosting does not train models in parallel: the base classifiers are trained in a sequence (Freund and Schapire, 1997; Schapire, 1990). Each subsequent base classifier is built to emphasize the training instances that previous learners misclassified. Boosting is a supervised search in the space of weak learners which may be regularized (see Chapters 9 and 14).

2.7 Data Dependencies and Computational Complexity

We are looking for patterns in the data: to extract the patterns, we analyze relationships between instances. We are interested in how one instance relates to other instances. Yet, not every pair of instances is of importance. Which data dependencies should we look at? How do dependencies influence computational time? These questions are crucial to understand why certain algorithms are favored on contemporary hardware, and they are equally important to see how quantum computers reduce computational complexity.

As a starting point, consider the trivial case: we compare every data instance with every other one. If the data instances are nodes in a graph, the dependencies form a complete graph K_N—this is an $N : N$ dependency. This situation frequently occurs in learning algorithms. For instance, if we calculate a distance matrix, we will have this type of dependency. The kernel matrix of a support vector machine (Chapter 7) also exhibits $N : N$ data dependency. In a distributed computing environment, $N : N$

dependencies will lead to excess communication between the nodes, as data instances will be located in remote nodes, and their feature vectors or other description must be exchanged to establish the distance.

Points that lie the furthest apart are not especially interesting to compare, but it is not immediately obvious which points lie close to one another in a high-dimensional space. Spatial data structures help in reducing the size of sets of data instances that are worth comparing. Building a tree-based spatial index often pays off. Examples include the R*-tree (Beckmann et al., 1990) or the X-tree (Berchtold et al., 1996) for data from a vector space, or the M-tree (Ciaccia et al., 1997) for data from a metric space. The height of such a tree-based index is $O(\log N)$ for a database of N objects in the worst case. Such structures not only reduce the necessary comparisons, but may also improve the performance of the learner, as in the case of clustering-based support vector machines (Section 7.9).

In many learning algorithms, data instances are never compared directly. Neural networks, for example, adjust their weights as data instances arrive at the input nodes (Chapter 6). The weights act as proxies; they capture relations between instances without directly comparing them. If there are K weights in total in a given topology of the network, the dependency pattern will be $N : K$. If $N \gg K$, it becomes clear why there are theoretical computational advantages to such a scheme. Under the same assumption, parallel architectures easily accelerate actual computations (Section 10.2).

Data dependencies constitute a large part of the computational complexity. If the data instances are regular dense vectors of d dimensions, calculating a distance matrix with $N : N$ dependencies will require $O(N^2 d)$ time complexity. If we use a tree-based spatial index, the run time is reduced to $O(dN \log N)$. With access to quantum memory, this complexity reduces to $O(\log \text{poly}(N))$—an exponential speedup over the classical case (Section 10.2).

If proxies are present to replace direct data dependencies, the time complexity will be in the range of $O(NK)$. The overhead of updating weights can outweigh the benefit of lower theoretical complexity.

Learning is an iterative process; hence, eventual computational complexity will depend on the form of optimization performed and on the speed of convergence. A vast body of work is devoted to reformulating the form of optimization in learning algorithms—some are more efficient than others. Restricting the algorithm often yields reduced complexity. For instance, support vector machines with linear kernels can be trained in linear time (Joachims, 2006).

Convergence is not always fast, and some algorithms never converge—in these cases, training stops after reaching appropriate conditions. The number of iterations is sometimes hard to predict.

In the broader picture, learning a classifier with a nonconvex loss function is an NP-hard problem even for simple classes of functions (Feldman et al., 2012)—this is the key reasoning behind using convex formulation for the optimization (Section 2.4). In some special cases, such as support vector machines, it pays off: direct optimization of a nonconvex objective function leads to higher accuracy and faster training (Collobert et al., 2006).

Quantum Mechanics

3

Quantum mechanics is a rich collection of theories that provide the most complete description of nature to date. Some aspects of it are notoriously hard to grasp, yet a tiny subset of concepts will be sufficient to understand the relationship between machine learning and quantum computing. This chapter collects these relevant concepts, and provides a brief introduction, but it deliberately omits important topics that are not crucial to understanding the rest of the book; for instance, we do not re-enumerate the postulates of quantum mechanics.

The mathematical toolkit resembles that of machine learning, albeit the context is different. We will rely on linear algebra, and, to a much lesser extent, on multivariate calculus. Unfortunately, the notation used by physicists differs from that in other applications of linear algebra. We use the standard quantum mechanical conventions for the notation, while attempting to keeping it in line with that used in the rest of the book.

We start this chapter by introducing the fundamental concept of the superposition of state, which will be crucial for all algorithms discussed later (Section 3.1). We follow this with an alternative formulation for states by density matrices, which is often more convenient to use (Section 3.2). Another phenomenon, entanglement, show stronger correlations than what classical systems can realize, and this is increasingly exploited in quantum computations (Section 3.3).

The evolution of closed quantum systems is linear and reversible, which has repercussions for learning algorithms (Section 3.4). Measurement on a quantum system, on the other hand, is strictly nonreversible, which makes it possible to introduce nonlinearity in certain algorithms (Section 3.5).

The uncertainty principle (Section 3.6) provides an explanation for quantum tunneling (Section 3.7), which in turn is useful in certain optimizations, particularly in ones that rely on the adiabatic theorem (Section 3.8).

The last section in this chapter gives a simple explanation of why arbitrary quantum states cannot be cloned, which makes copying of quantum data impossible (Section 3.9).

This chapter focuses on concepts that are common to quantum computing and derived learning algorithms. Additional concepts—such as representation theory—will be introduced in chapters where they are relevant.

Quantum Machine Learning. http://dx.doi.org/10.1016/B978-0-12-800953-6.00001-1

3.1 States and Superposition

The state in quantum physics contains statistical information about a quantum system. Mathematically, it is represented by a vector—the state vector. A state is essentially a probability density; thus, it does not directly describe physical quantities such as mass or charge density.

The state vector is an element of a Hilbert space. The choice of Hilbert space depends on the purpose, but in quantum information theory, it is most often $\mathbb{C}^n$. A vector has a special notation in quantum mechanics, the Dirac notation. A vector—also called a ket—is denoted by

$$|\psi\rangle, \tag{3.1}$$

where ψ is just a label. This label is as arbitrary as the name of a vector variable in other applications of linear algebra; for instance, the $\mathbf{x}_i$ data instances in Chapter 2 could be denoted by any other character.

The ket notation abstracts the vector space: it no longer matters whether it is a finite-dimensional complex space or the infinite-dimensional space of Lebesgue square-integrable functions. When the ket is in finite dimensions, it is a column vector.

Since the state vectors are related to probabilities, some form of normalization must be imposed on the vectors. In a general Hilbert space setting, we require the norm of the state vectors to equal 1:

$$\| |\psi\rangle \| = 1. \tag{3.2}$$

While it appears logical to denote the zero vector by $|0\rangle$, this notation is reserved for a vector in the computational basis (Section 4.1). The null vector will be denoted by 0.

The dual of a ket is a bra. Mathematically, a bra is the conjugate transpose of a ket:

$$\langle\psi| = |\psi\rangle^\dagger. \tag{3.3}$$

If the Hilbert space is a finite-dimensional real or complex space, a bra corresponds to a row vector. With this notation, an inner product between two states $|\phi\rangle$ and $|\psi\rangle$ becomes

$$\langle\phi|\psi\rangle. \tag{3.4}$$

If we choose a basis $\{|k_i\rangle\}$ in the Hilbert space of the quantum system, then a state vector $|\psi\rangle$ expands as the linear combination of the basis vectors:

$$|\psi\rangle = \sum_i \alpha_i |k_i\rangle, \tag{3.5}$$

where the α_i coefficients are complex numbers, and the sum may be infinite, depending on the dimensions of the Hilbert space. The α_i coefficients are called probability amplitudes, and the normalization constraint on the state vector implies that

$$\sum_i |\alpha_i|^2 = 1. \tag{3.6}$$

The sum in Equation 3.5 is called a quantum superposition of the states $|k_i\rangle$. Any sum of state vectors is a superposition, subject to renormalization.

The superposition of a quantum system expresses that the system exists in all of its theoretically possible states simultaneously. When a measurement is performed, however, only one result is obtained, with a probability proportional to the weight of the corresponding vector in the linear combination (Section 3.5).

3.2 Density Matrix Representation and Mixed States

An alternative representation of states is by density matrices. They are also called density operators; we use the two terms interchangeably. The density matrix is an operator formed by the outer product of a state vector:

$$\rho = |\psi\rangle\langle\psi|. \tag{3.7}$$

A state ρ that can be written in this form is called a pure state. The state vector might be in a superposition, but the corresponding density matrix will still describe a pure state.

Since quantum physics is quintessentially probabilistic, it is advantageous to think of a pure state as a pure ensemble, a collection of identical particles with the same physical configuration. A pure ensemble is described by one state function ψ for all its particles. The following properties hold for pure states:

- A density matrix is idempotent: $\rho^2 = |\psi\rangle\langle\psi|\psi\rangle\langle\psi| = |\psi\rangle\langle\psi| = \rho$.
- Given any orthonormal basis $\{|n\rangle\}$, the trace of a density matrix is 1: $\text{tr}(\rho) = \sum_n \langle n|\rho|n\rangle = \sum_n \langle n|\psi\rangle\langle\psi|n\rangle = \sum_n \langle\psi|n\rangle\langle n|\psi\rangle = 1$.
- Similarly, $\text{tr}(\rho^2) = 1$.
- Hermiticity: $\rho^\dagger = (|\psi\rangle\langle\psi|)^\dagger = |\psi\rangle\langle\psi| = \rho$.
- Positive semidefinite: $\langle\phi|\rho|\phi\rangle = \langle\phi|\psi\rangle\langle\psi|\phi\rangle = |\langle\phi|\psi\rangle|^2 \geq 0$.

Density matrices allow for states of another type, mixed states. A mixed state is a mixture of projectors onto pure states:

$$\rho_{\text{mixed}} = \sum_i p_i |\psi_i\rangle\langle\psi_i|. \tag{3.8}$$

Taking a statistical interpretation again, a mixed state consists of identical particles, but portions are in different physical configurations. A mixture is described by a set of states ψ_i with corresponding probabilities. This justifies the name density matrix: a mixed state is a distribution over pure states. The properties of a mixed state are as follows:

- Idempotency is violated: $\rho_{\text{mixed}}^2 = \left(\sum_i p_i|\psi_i\rangle\langle\psi_i|\right)\left(\sum_j p_j|\psi_j\rangle\langle\psi_j|\right)$
  $= \sum_{i,j} p_i p_j |\psi_i\rangle\langle\psi_i|\psi_j\rangle\langle\psi_j| \neq \rho_{\text{mixed}}$.
- $\text{tr}(\rho_{\text{mixed}}) = 1$.
- However, $\text{tr}(\rho_{\text{mixed}}^2) < 1$.

- Hermicity.
- Positive semidefinite.

We do not normally denote mixed states with a lower index as above; instead, we write ρ for both mixed and pure states.

To highlight the distinction between superposition and mixed states, fix a basis $\{|0\rangle, |1\rangle\}$. A superposition in this two-dimensional space is a sum of two vectors:

$$|\psi\rangle = \alpha|0\rangle + \beta|1\rangle, \tag{3.9}$$

where $|\alpha|^2 + |\beta|^2 = 1$. The corresponding density matrix is

$$\rho = \begin{pmatrix} |\alpha|^2 & \alpha\beta^* \\ \alpha^*\beta & |\beta|^2 \end{pmatrix}, \tag{3.10}$$

where * stands for complex conjugation.

A mixed state is, on the other hand, a sum of projectors:

$$\rho_{\text{mixed}} = |\alpha|^2|0\rangle\langle 0| + |\beta|^2|1\rangle\langle 1| = \begin{pmatrix} |\alpha|^2 & 0 \\ 0 & |\beta|^2 \end{pmatrix}. \tag{3.11}$$

Interference terms—the off-diagonal elements—are present in the density matrix of a pure state (Equation 3.10), but they are absent in a mixed state (Equation 3.11).

A density matrix is basis-dependent, but the trace of it is invariant with respect to a transformation of basis.

The density matrix of a state is not unique. Different superpositions may have the same density matrix:

$$|\psi_1\rangle = \frac{1}{\sqrt{2}}|0\rangle + \frac{1}{\sqrt{2}}\left(\frac{1}{\sqrt{2}}(|0\rangle - |1\rangle)\right), \tag{3.12}$$

$$|\psi_2\rangle = \frac{1}{\sqrt[4]{3}}\left(\frac{\sqrt{3}}{2}|0\rangle - \frac{1}{2}|1\rangle\right) + \frac{3}{4}(1 - \frac{1}{\sqrt{3}})|0\rangle + \frac{1}{4}(1 - \frac{1}{\sqrt{3}})|1\rangle, \tag{3.13}$$

$$\rho_1 = |\psi_1\rangle\langle\psi_1| = \begin{pmatrix} \frac{3}{4} & -\frac{1}{4} \\ -\frac{1}{4} & \frac{1}{4} \end{pmatrix} = |\psi_2\rangle\langle\psi_2| = \rho_2. \tag{3.14}$$

An infinite number of ensembles can generate the same density matrix. Absorb the probability of a state vector in the vector itself: $|\tilde{\psi}\rangle = \alpha_i|\psi\rangle$. Then the following is true:

Theorem 3.1. *The sets $\{|\tilde{\psi}\rangle_i\}$ and $\{|\tilde{\phi}_j\rangle\}$ will have the same density matrix if and only if $|\tilde{\psi}_i\rangle = \sum_j u_{ij}|\tilde{\phi}_j\rangle$, where the u_{ij} elements form a unitary transformation.*

While there is a clear loss of information by not having a one-to-one correspondence with state vectors, density matrices provide an elegant description of probabilities, and they are often preferred over the state vector formalism.

3.3 Composite Systems and Entanglement

Not every collection of particles is a pure state or a mixed state. Composite quantum systems are made up of two or more distinct physical systems. Unlike in classical physics, particles can become coupled or entangled, making the composite system more than the sum of the components.

The state space of a composite system is the tensor product of the states of the component physical systems. For instance, for two components A and B, the total Hilbert space of the composite system becomes $\mathcal{H}_{AB} = \mathcal{H}_A \otimes \mathcal{H}_B$. A state vector on the composite space is written as $|\psi\rangle_{AB} = |\psi\rangle_A \otimes |\psi\rangle_B$. The tensor product is often abbreviated as $|\psi\rangle_A|\psi\rangle_B$, or, equivalently, the labels are written in the same ket $|\psi_A\psi_B\rangle$.

As an example, assume that the component spaces are two-dimensional, and choose a basis in each. Then, a tensor product of two states yields the following composite state:

$$\begin{pmatrix} \alpha \\ \beta \end{pmatrix}_A \otimes \begin{pmatrix} \gamma \\ \delta \end{pmatrix}_B = \begin{pmatrix} \alpha\gamma \\ \alpha\delta \\ \beta\gamma \\ \beta\delta \end{pmatrix}_{AB}. \tag{3.15}$$

The Schmidt decomposition allows a way of expanding a vector on a product space. The following is true:

Theorem 3.2. *Let $\mathcal{H}_A$ and $\mathcal{H}_B$ have orthonormal bases $\{e_1, e_2, \ldots, e_n\}$ and $\{f_1, f_2, \ldots, f_m\}$, respectively. Then, any bipartite state $|\psi\rangle$ on $\mathcal{H}_A \otimes \mathcal{H}_B$ can be written as*

$$|\psi\rangle = \sum_{k=1}^{r} a_k |e_k\rangle \otimes |f_n\rangle, \tag{3.16}$$

where r is the Schmidt rank.

This decomposition resembles singular value decomposition.

The density matrix representation is useful for the description of individual subsystems of a composite quantum system. The density matrix of the composite system is provided by a tensor product:

$$\rho_{AB} = \rho_A \otimes \rho_B. \tag{3.17}$$

We can recover the components ρ_A and ρ_B from the composite vector ρ_{AB}. A subsystem of a composite quantum system is described by a reduced density matrix. The reduced density matrix for a subsystem A is defined by

$$\rho_A = \mathrm{tr}_B(\rho_{AB}), \tag{3.18}$$

where tr_B is a partial trace operator over system B. In taking the partial trace, the probability amplitudes belonging to system B vanish—this is due to $\mathrm{tr}(\rho) = 1$ for any density matrix. This procedure is also called "tracing out." Only the amplitudes belonging to system A remain.

Density matrices and the partial trace operator allow us to find the rank of a Schmidt decomposition. Take an orthonormal basis $\{|f_k\rangle\}$ in system B. Then, the reduced density matrix ρ_A is

$$\rho_A = \text{tr}_B(\rho_{AB}) = \sum_k \langle f_k|\rho_{AB}|f_k\rangle. \tag{3.19}$$

On the other hand, we can write

$$\rho_A = \sum_i \langle f_i| \left(\sum_{k=1}^r a_k|e_k\rangle|f_i\rangle\right)\left(\sum_{k=1}^r a_k\langle e_k|\langle f_i|\right)|f_i\rangle = \sum_{k=1}^r |a_k|^2|e_k\rangle\langle e_k|. \tag{3.20}$$

Hence, we get rank(ρ_A) =Schmidt rank of ρ_{AB}.

Let us study state vectors on the Hilbert space $\mathcal{H}_{AB}$. For example, given a basis $\{|0\rangle, |1\rangle\}$, the most general pure state is given as

$$\psi\rangle = \alpha|00\rangle + \beta|01\rangle + \gamma|10\rangle + \delta|11\rangle. \tag{3.21}$$

Take as an example a Bell state, defined as $|\phi^+\rangle = \frac{|00\rangle+|11\rangle}{\sqrt{2}}$ (Section 4.1). This state cannot be written as a product of two states.

Suppose there are states $\alpha|0\rangle + \beta|1\rangle$ and $\gamma|0\rangle + \delta|1\rangle$:

$$\frac{|00\rangle + |11\rangle}{\sqrt{2}} = (\alpha|0\rangle + \beta|1\rangle) \otimes (\gamma|0\rangle + \delta|1\rangle). \tag{3.22}$$

Then,

$$(\alpha|0\rangle + \beta|1\rangle) \otimes (\gamma|0\rangle + \delta|1\rangle) = \alpha\gamma|00\rangle + \alpha\delta|01\rangle + \beta\gamma|10\rangle + \beta\delta|11\rangle. \tag{3.23}$$

That is, we need to find a solution to the equations

$$\alpha\gamma = \frac{1}{\sqrt{2}}, \qquad \alpha\delta = 0, \qquad \beta\gamma = 0, \qquad \beta\delta = \frac{1}{\sqrt{2}}. \tag{3.24}$$

This system of equations does not have a solution; therefore, $\frac{|00\rangle+|11\rangle}{\sqrt{2}}$ cannot be a product state.

Composite states that can be written as a product state are called separable, whereas other composite states are entangled.

Density matrices reveal information about entangled states. This Bell state has the density operator

$$\rho_{AB} = \frac{|00\rangle + |11\rangle}{\sqrt{2}} \frac{\langle 00| + \langle 11|}{\sqrt{2}} \tag{3.25}$$

$$= \frac{|00\rangle\langle 00| + |11\rangle\langle 00| + |00\rangle\langle 11| + |11\rangle\langle 11|}{2}. \tag{3.26}$$

Tracing it out in system B, we get

$$\rho_A = \text{tr}_B(\rho_{AB}) \tag{3.27}$$

$$= \frac{\text{tr}_B(|00\rangle\langle 00|) + \text{tr}_B(|11\rangle\langle 00|) + \text{tr}_B(|00\rangle\langle 11|) + \text{tr}_B(|11\rangle\langle 11|)}{2} \tag{3.28}$$

$$= \frac{|0\rangle\langle 0|\langle 0|0\rangle + |1\rangle\langle 0|\langle 0|1\rangle + |0\rangle\langle 1|\langle 1|0\rangle + |1\rangle\langle 1|\langle 1|1\rangle}{2} \tag{3.29}$$

$$= \frac{|0\rangle\langle 0| + |1\rangle\langle 1|}{2} = \frac{I}{2}. \tag{3.30}$$

The trace of this is less than 1; therefore, it is a mixed state. The entangled state is pure in the composite space, but surprisingly, its partial traces are mixed states. From the Schmidt decomposition, we see that a state is entangled if and only if its Schmidt rank is strictly greater than 1. Therefore, a bipartite pure state is entangled if and only if its reduced states are mixed states.

The reverse process is called purification: given a mixed state, we are interested in finding a pure state that gives the mixed state as its reduced density matrix. The following theorem holds:

Theorem 3.3. *Let ρ_A be a density matrix acting on a Hilbert space $\mathcal{H}_A$ of finite dimension n. Then, there exists a Hilbert space $\mathcal{H}_B$ and a pure state $|\psi\rangle \in \mathcal{H}_A \otimes \mathcal{H}_B$ such that the partial trace of $|\psi\rangle\langle\psi|$ with respect to $\mathcal{H}_B$ equals ρ_A:*

$$\mathrm{tr}_B\left(|\psi\rangle\langle\psi|\right) = \rho_A. \tag{3.31}$$

The pure state $|\psi\rangle$ is the purification of ρ_A.

The purification is not unique; there are many pure states that reduce to the same density matrix. We call two states maximally entangled if the reduced density matrix is diagonal with equal probabilities as entries.

Density matrices are able to prove the presence of entanglement in other forms. The Peres-Horodecki criterion is a necessary condition for the density matrix of a composite system to be separable. For two- or three-dimensional cases, it is a sufficient condition (Horodecki et al., 1996). It is useful for mixed states, where the Schmidt decomposition does not apply.

Assume a general state ρ_{AB} acts on a composite Hilbert space

$$\rho_{AB} = \sum_{ijkl} p^{ij}_{kl} |i\rangle\langle j| \otimes |k\rangle\langle l|. \tag{3.32}$$

Its partial transpose with respect to system B is defined as

$$\rho_{AB}^{\top_B} = \sum_{ijkl} p^{ij}_{kl} |i\rangle\langle j| \otimes \left(|k\rangle\langle l|\right)^{\top} = \sum_{ijkl} p^{ij}_{kl} |i\rangle\langle j| \otimes |l\rangle\langle k|. \tag{3.33}$$

If ρ_{AB} is separable, $\rho_{AB}^{\top_B}$ has nonnegative eigenvalues.

Quantum entanglement has been experimentally verified (Aspect et al., 1982); it is not just an abstract mathematical concept, it is an aspect of reality. Entanglement is a correlation between two systems that is stronger than what classical systems are able to produce. A local hidden variable theory is one in which distant events do not have an instantaneous effect on local ones—seemingly instantaneous events can always be explained by hidden variables in the system. Entanglement may produce instantaneous correlations between remote systems which cannot be explained by local hidden variable theories; this phenomenon is called nonlocality. Classical systems cannot produce nonlocal phenomena.

Bell's theorem draws an important line between quantum and classical correlations of composite systems (Bell, 1964). The limit is easy to test when given in the following inequality (the Clauser-Horne-Shimony-Holt inequality; Clauser et al., 1969):

$$\mathbf{C}[A(a), B(b)] + \mathbf{C}[A(a), B(b')] + \mathbf{C}[A(a'), B(b)] - \mathbf{C}[A(a'), B(b')] \leq 2, \quad (3.34)$$

where a and a' are detector settings on side A of the composite system, b and b' are detector settings on side B, and $\mathbf{C}$ denotes correlation. This is a sharp limit: any correlation violating this inequality is nonlocal.

Entanglement and nonlocality are not the same, however. Entanglement is a necessary condition for nonlocality, but more entanglement does not mean more nonlocality (Vidick and Wehner, 2011). Nonlocality is a more generic term: "There exist in nature channels connecting two (or more) distant partners, that can distribute correlations which can neither be caused by the exchange of a signal (the channel does not allow signalling, and moreover, a hypothetical signal should travel faster than light), nor be due to predetermined agreement..." (Scarani, 2006).

Entanglement is a powerful resource that is often exploited in quantum computing and quantum information theory. This is reflected by the cost of simulating entanglement by classical composite systems: exponentially more communication is necessary between the component systems (Brassard et al., 1999).

3.4 Evolution

Unobserved, a quantum mechanical system evolves continuously and deterministically. This is in sharp contrast with the unpredictable jumps that occur during a measurement (Section 3.5). The evolution is described by the Schrödinger equation.

In its most general form, the Schrödinger equation reads as follows:

$$i\hbar\frac{\partial}{\partial t}|\psi\rangle = H|\psi\rangle, \quad (3.35)$$

where H is the Hamiltonian operator, and $\hbar$ is Planck's constant—its actual value is not important to us. The Hamiltonian characterizes the total energy of a system and takes different forms depending on the situation.

In this context, the $|\psi\rangle$ state vector is also called the wave function of the quantum system. The wave function nomenclature justifies the abstraction level of the bra-ket notation: as mentioned in Section 3.1, a ket is simply a vector in a Hilbert space. If we think about the state as a wave function, this often implies that it is an actual function, an element of the infinite-dimensional Hilbert space of Lebesgue square-integrable functions. We, however, almost always use a finite-dimensional complex vector space as the underlying Hilbert space. A notable exception is quantum tunneling, where the wave function has additional explanatory meaning (Section 3.7). In turn, quantum annealing relies on quantum tunneling (Section 14.1); hence, it is worth taking note of the function space interpretation.

An equivalent way of writing the Schrödinger equation is with density matrices:

$$i\hbar\frac{\partial\rho}{\partial t} = [H,\rho], \tag{3.36}$$

where [,] is the commutator operator: $[H,\rho] = H\rho - \rho H$.

The Hamiltonian is a Hermitian operator; therefore, it has a spectral decomposition. If the Hamiltonian is independent of time, the following equation gives the time-independent Schrödinger equation for the state vector:

$$E|\psi\rangle = H|\psi\rangle, \tag{3.37}$$

where E is the energy of the state, which is an eigenvalue of the Hamiltonian. Solving this equation yields the stationary states for a system—these are also called energy eigenstates. If we understand these states, solving the time-dependent Schrödinger equation becomes easier for any other state. The smallest eigenvalue is called the ground-state energy, which has a special role in many applications, including adiabatic quantum computing, where an adiabatic change of the ground state will yield the optimum of a function being studied (Section 3.8 and Chapter 14). An excited state is any state with energy greater than the ground state.

Consider an eigenstate ψ_α of the Hamiltonian $H\psi_\alpha = E_\alpha\psi_\alpha$. Taking the Taylor expansion of the exponential, we observe how the time evolution operator acts on this eigenstate:

$$e^{-iHt/\hbar}|\psi_\alpha(0)\rangle \tag{3.38}$$

$$= \left(\sum_n \frac{1}{n!}\left(\frac{H}{i\hbar}\right)^n t^n\right)|\psi_\alpha(0)\rangle \tag{3.39}$$

$$= \left(\sum_n \frac{1}{n!}\left(\frac{E_\alpha}{i\hbar}\right)^n t^n\right)|\psi_\alpha(0)\rangle \tag{3.40}$$

$$= e^{-iE_\alpha t/\hbar}|\psi_\alpha(0)\rangle. \tag{3.41}$$

We define $U(H,t) = e^{-iHt/\hbar}$. This is the time evolution operator of a closed quantum system. It is a unitary operator, and this property is why quantum gates are reversible. The intrinsically unitary nature of quantum systems has important implications for learning algorithms using quantum hardware. We often denote $U(H,t)$ by the single letter U if the Hamiltonian is understood or is not important, and we imply time dependency implicitly.

The evolution in the density matrix representation reads

$$\rho \mapsto U\rho U^\dagger. \tag{3.42}$$

U is a linear operator, so it acts independently on each term of a superposition. The state is a superposition of its eigenstates, and thus its time evolution is given by

$$|\psi(t)\rangle = \sum_\alpha c_\alpha e^{-iE_\alpha t/\hbar}|\psi_\alpha(0)\rangle. \tag{3.43}$$

The time evolution operator, being unitary, preserves the l_2 norm of the state—that is, the probability amplitude will sum to 1 at every time step. This result means even more: U does not change the probabilities of the eigenstates, but only changes the phases.

The matrix form of U depends on the basis. If we take any orthonormal basis, elements of the time evolution matrix acquire a clear physical meaning as the transition amplitudes between the corresponding eigenstates of this basis (Fayngold and Fayngold, 2013, p. 297). The transition amplitudes are generally time-dependent.

The unitary evolution reveals insights into the nomenclature "probability amplitudes." The norm of the state vector is 1, and the components of the norm are constant. The probability amplitudes, however, oscillate between time steps: their phase changes.

A second look at Equation 3.41 reveals that an eigenvector of the Hamiltonian is an eigenvector of the time evolution operator. The eigenvalue is a complex exponential, which means U is not Hermitian.

3.5 Measurement

The state vector evolves deterministically as the continuous solution of the wave equation. All the while, the state vector is in a superposition of component states. What happens to a superposition when we perform a measurement on the system?

Before we can attempt to answer that question, we must pay attention to an equally important one: What is being measured? It is the probability amplitude that evolves in a deterministic manner, and not a measurable characteristic of the system (Fayngold and Fayngold, 2013, p. 558).

An observable quantity, such as the energy or momentum of a particle, is associated with a mathematical operator, the observable. The observable is a Hermitian operator acting on the state space M with spectral decomposition

$$M = \sum_i \alpha_i P_i, \tag{3.44}$$

where P_i is a projector onto the eigenspace of the operator M with eigenvalue α_i. In other words, an observable is a weighted sum of projectors. The possible outcomes of the measurement correspond to the eigenvalues α_i. Since M is Hermitian, the eigenvalues are real.

The projectors are idempotent by definition, and they map to the eigenspace of the operator, they are orthogonal, and their sum is the identity:

$$P_i P_j = \delta_{ij} P_i, \tag{3.45}$$

$$\sum_i P_i = I. \tag{3.46}$$

Returning to the original question, we find a discontinuity occurs when a measurement is performed. All components of the superposition vanish, except one. We observe an eigenvalue α_i with the following probability:

$$\mathbf{P}(\alpha_i) = \langle \psi | P_i | \psi \rangle. \tag{3.47}$$

Thus, the outcome of a measurement is inherently probabilistic. This formula is also called Born's rule. The system will be in the following state immediately after measurement:

$$\frac{P_i | \psi \rangle}{\sqrt{\langle \psi | P_i | \psi \rangle}}. \tag{3.48}$$

This procedure is called a projective or von Neumann measurement. The measurement is irreversible and causes loss of information, as we cannot learn more about the superposition before the measurement. A simple explanation for the phenomenon is that the discontinuity arises from the interaction of a classical measuring instrument and the quantum system. More elaborate explanations abound, but they are not relevant for the rest of the discussion.

The loss of information from a quantum system is also called decoherence. As the quantum system interacts with its environment—for instance, with the measuring instrument—components of the state vector are decoupled from a coherent system, and entangle with the surroundings. A global state vector of the system and the environment remains coherent: it is only the system we are observing that loses coherence. Hence, decoherence does not explain the discontinuity of the measurement, it only explains why an observer no longer sees the superposition. Furthermore, decoherence occurs spontaneously between the environment and the quantum system even if we do not perform a measurement. This makes the realization of quantum computing a tough challenge, as a quantum computer relies on the undisturbed evolution of quantum superpositions.

Measurements with the density matrix representation mirror the projective measurement scheme. The probability of obtaining an output α_i is

$$\mathbf{P}(\alpha_i) = \langle \psi | P_i | \psi \rangle = \langle \psi | P_i P_i | \psi \rangle \tag{3.49}$$

$$= \langle \psi | P_i^\dagger P_i | \psi \rangle = \mathrm{tr}(P_i^\dagger P_i | \psi \rangle \langle \psi |) = \mathrm{tr}(P_i^\dagger P_i \rho). \tag{3.50}$$

The density matrix after measurement must be renormalized:

$$\frac{P_i \rho P_i^\dagger}{\mathrm{tr}(P_i^\dagger P_i \rho)}. \tag{3.51}$$

Projective measurements are restricted to orthogonal states: they project on an eigenspace. We may want to observe outputs that do not belong to orthogonal states. Positive operator–valued measures (POVMs) overcome this limitation.

A POVM is set of positive Hermitian operators $\{P_i\}$ that satisfies the completeness relation:

$$\sum_i P_i = I. \tag{3.52}$$

That is, the orthogonality and idempotency constraints of projective measurements are relaxed. Thus, a POVM represents a nonoptimal measurement that is not designed to return an eigenstate of the system. Instead, a POVM measures along unit vectors that are not orthogonal.

The probability of obtaining an output α_i is given by a formula similar to that for projective measurements:

$$\mathbf{P}(\alpha_i) = \langle\psi|P_i|\psi\rangle. \tag{3.53}$$

Similarly, the system will have to be renormalized after measurement:

$$\frac{P_i|\psi\rangle}{\sqrt{\langle\psi|P_i|\psi\rangle}}. \tag{3.54}$$

We may reduce a POVM to a projective measurement on a larger Hilbert space. We couple the original system with another system called the ancilla (Fayngold and Fayngold, 2013, p. 660). We let the joint system evolve until the nonorthogonal unit vectors corresponding to outputs become orthogonal. In this larger Hilbert space, the POVM reduces to a projective measurement. This is a common pattern in many applications of quantum information theory: ancilla systems aid understanding or implementing a specific target easier.

3.6 Uncertainty Relations

If two observables do not commute, a state cannot be a simultaneous eigenvector of both in general (Cohen-Tannoudji et al., 1996, p. 233). This leads to a form of the uncertainty relation similar to the one found by Heisenberg in his analysis of sequential measurements of position and momentum. This original relation states that there is a fundamental limit to the precision with which the position and momentum of a particle can be known.

The expectation value of an observable A—a Hermitian operator—is $\langle A\rangle = \langle\psi|A|\psi\rangle$. Its standard deviation is $\sigma_A = \sqrt{\langle A^2\rangle - \langle A\rangle^2}$. In the most general form, the uncertainty principle is given by

$$\sigma_A\sigma_B \geq \frac{1}{2}\left|\langle[A,B]\rangle\right|. \tag{3.55}$$

This relation clearly shows that uncertainty emerges from the noncommutativity of the operators. It implies that the observables are incompatible in a physical setting. The incompatibility is unrelated to subsequent measurements in a single experiment. Rather, it means that preparing identical states, $|\psi\rangle$, we measure one observable in one subset, and the other observable in the other subset. In this case, the standard deviation of the measurements will satisfy the inequality in Equation 3.6.

As long as two operators do not commute, they will be subjected to a corresponding uncertainty principle. This attracted attention from other communities who apply quantum-like observables to describe phenomena, for instance, in cognitive science (Pothos and Busemeyer, 2013).

Interestingly, the uncertainty principle implies nonlocality (Oppenheim and Wehner, 2010). The uncertainty principle is a restriction on measurements made on a single system, and nonlocality is a restriction on measurements conducted on two systems. Yet, by treating both nonlocality and uncertainty as a coding problem, we find these restrictions are related.

3.7 Tunneling

Quantum tunneling is a phenomenon that has no classical counterpart, but that has proved to be tremendously successful in applications. In a machine learning context, it plays a part in quantum annealing (Section 14.1).

In a classical system, assume that a moving object has a mechanical energy E, and it reaches a barrier with potential energy $U(x)$. If $E < U(x)$, the object will not be able to surmount the barrier. In quantum systems, there is such a possibility.

A quantum particle may appear on either side of the barrier—after all, its location is probabilistic. The introduction of a potential barrier merely changes the distribution. A particle with definite $E < U(x)$, and even $E \ll U(x)$, can pass through the barrier, almost as if there were a "tunnel" cut across. As the potential barrier increases, the probability of tunneling decreases: taller and wider barriers will see fewer particles tunneling through.

Let us denote the two turning points on the two sides of the barrier by x_1 and x_2 (Figure 3.1). The region $x_1 < x < x_2$ is classically inaccessible. From the position indeterminancy $\Delta x \leq |x_2 - x_1|$, it follows that, as $\Delta p \geq \hbar/2\Delta x$, the particle has an indeterminancy in its energy: $E \pm \frac{1}{2}\Delta E$. Thus, the particle's energy may exceed the barrier, and the particle passes over the barrier.

To understand why tunneling occurs, the wave nature of particles helps. The process resembles the total internal reflection of a wave in classical physics (Fayngold and Fayngold, 2013, p. 241). The total reflection does not happen exactly at the interface between the two media: it is preceded by partial penetration of the wave into the second layer. If the second layer has finite width, the penetrated part of the wave partially leaks out on the other side of the layer, leading to frustrated internal reflection, and causing transmission.

3.8 Adiabatic Theorem

The adiabatic theorem, originally proved in Born and Fock (1928), has important applications in quantum computing (Section 4.3 and Chapter 14). An adiabatic process changes conditions gradually so as to allow the system to adapt its configuration. If the

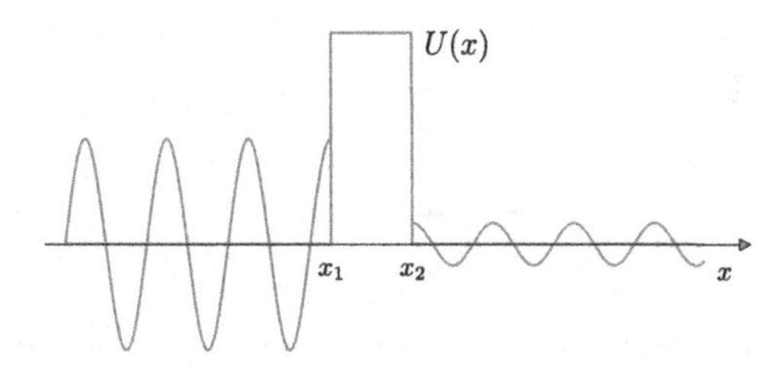

Figure 3.1 Quantum tunneling through a potential barrier between the classically inaccessible region $x_1 < x < x_2$. The particle that tunnels through has a decreased amplitude, but the same energy.

system starts in the ground state of an initial Hamiltonian, after the adiabatic change, it will end in the ground state of the final Hamiltonian.

Consider a Hamiltonian H_0 whose unique ground state can be easily constructed, and another Hamiltonian H_1 whose ground-state energy we are interested in. To find this ground state, one considers the Hamiltonian

$$H(\lambda) = (1 - \lambda)H_0 + \lambda H_1, \tag{3.56}$$

with $\lambda = \lambda(t)$ as a time-dependent parameter with a range in $[0, 1]$. The quantum adiabatic theorem states that if we start in the ground state of $H(0) = H_0$ and we slowly increase the parameter λ, then the system will evolve to the ground state of H_1 if $H(\lambda)$ is always gapped—that is, there is no degeneracy for the ground-state energy (Figure 3.2). The adiabatic theorem also states that the correctness of the change to the system depends critically on the time $t_1 - t_0$ during which the change takes place. This should be large enough, and it depends on the minimum gap between the ground state and the first excited state throughout the adiabatic process. While this is intuitive, a gapless version of theorem also exists, in which case the duration must be estimated in some other way (Avron and Elgart, 1999).

3.9 No-Cloning Theorem

Cloning of a pure state $|\psi\rangle$ means a procedure with a separable state as an output: $|\psi\rangle \otimes |\psi\rangle$.

We start by adding an ancilla system with the same state space, but the initial state is unrelated to the state being cloned: $|\psi\rangle \otimes |0\rangle$. Cloning thus means finding a unitary operator such that it evolves this initial state to the desired output:

$$U\left(|\psi\rangle \otimes |0\rangle\right) = |\psi\rangle \otimes |\psi\rangle. \tag{3.57}$$

Since cloning should work for any state in the state space, U has a similar effect on a different state ϕ. Consider the inner product of ψ and ϕ together with the ancilla in the product space:

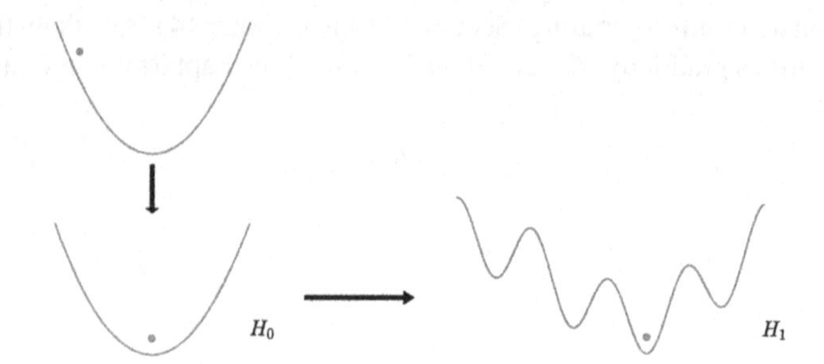

Figure 3.2 Using the adiabatic theorem, we reach the ground-state energy of a simple Hamiltonian H_0, then we gradually change the Hamiltonian until we reach the target H_1. If enough time is allowed for the change, the system remains in the ground state throughout the process.

$$\begin{aligned}\langle 0|\langle\phi||\psi\rangle|0\rangle &= \langle 0|\langle\phi|U^{\dagger}U|\psi\rangle|0\rangle \\ &= \langle\phi|\langle\phi|\psi\rangle|\psi\rangle. \end{aligned} \tag{3.58}$$

Thus, $\langle\phi|\psi\rangle = \langle\phi|\psi\rangle^2$. Since states are unit vectors, this implies that either $\langle\phi|\psi\rangle = 1$ or $\langle\phi|\psi\rangle = 0$. Yet, we required the two vectors to be arbitrary in the state space, so cloning is feasible only for specific cases. A unitary operator cannot clone a general quantum state.

The principle extends to mixed states, where it is known as the no-broadcast theorem.

The no-cloning theorem has important consequences for quantum computing: exact copying of data is not possible. Hence, the basic units of classical computing, such as a random access memory, require alternative concepts.

Quantum Computing

4

A quantum system makes a surprisingly efficient computer: a quantum algorithm may provide quadratic or even exponential speedup over the best known classical counterparts. Since learning problems involve tedious and computationally intensive calculations, this alone is a good reason to look at how quantum computers accelerate optimization.

This chapter looks at the most important concepts, and we restrict ourselves to theoretical constructs: we are not concerned with the physical implementation of quantum computers.

A qubit is the fundamental building block of quantum computing, and is a two-level quantum state (Section 4.1). By defining operations on single or multiple qubits, we can construct quantum circuits, which is one model of quantum computing (Section 4.2). An alternative model is adiabatic quantum computing, which manipulates states at a lower level of abstraction, using Hamiltonians (Section 4.3).

Irrespective of the computational model, one of the advantages of quantum computing is quantum parallelism—an application of superposition (Section 4.4). Learning algorithms rely on this parallelism for a speedup either directly or through Grover's search (Section 4.5). We highlight that the parallelism provided by quantum computers does not extend the range of calculable problems—some quintessential limits from classical computing still apply (Section 4.6).

We close this chapter by mentioning a few concepts of information theory (Section 4.7). The fidelity of states will be particularly useful, as it allows the definition of distance functions that are often used in machine learning.

4.1 Qubits and the Bloch Sphere

Any two-level quantum system can form a qubit—for example, polarized photons, spin-1/2 particles, excited atoms, and atoms in ground state.

A convenient choice of basis is $\{|0\rangle, |1\rangle\}$—this is called the computational basis. The general pure state of a qubit in this basis is $|\psi\rangle = \alpha|0\rangle + \beta|1\rangle$. Since $|\alpha|^2 + |\beta|^2| = 1$, we write

$$|\psi\rangle = e^{i\gamma}\left(\cos\frac{\theta}{2}|0\rangle + e^{i\phi}\sin\frac{\theta}{2}|1\rangle\right), \tag{4.1}$$

Quantum Machine Learning. http://dx.doi.org/10.1016/B978-0-12-800953-6.00001-1

with $0 \leq \theta \leq \pi$, $0 \leq \phi \leq 2\pi$. The global phase factor $e^{i\gamma}$ has no observable effects; hence, the formula reduces to

$$|\psi\rangle = \cos\frac{\theta}{2}|0\rangle + e^{i\phi}\sin\frac{\theta}{2}|1\rangle. \tag{4.2}$$

With the constraints on θ and ϕ, these two numbers define a point on the surface of the unit sphere in three dimensions. This sphere is called the Bloch sphere. Its purpose is to give a geometric explanation to single-qubit operations (Figure 4.1).

The Pauli matrices are a set of three 2×2 complex matrices which are Hermitian and unitary. They are

$$\sigma_x = \begin{pmatrix} 0 & 1 \\ 1 & 0 \end{pmatrix}, \quad \sigma_y = \begin{pmatrix} 0 & -i \\ i & 0 \end{pmatrix}, \quad \sigma_z = \begin{pmatrix} 1 & 0 \\ 0 & -1 \end{pmatrix}. \tag{4.3}$$

Together with the identity matrix I, they form a basis for the real Hilbert space of 2×2 complex Hermitian matrices. Each Pauli matrix is related to an operator that corresponds to an observable describing the spin of a spin-$1/2$ particle, in each of the corresponding three spatial directions.

In the density matrix representation, for pure states, we have

$$\rho = |\psi\rangle\langle\psi| = \begin{pmatrix} \cos^2\frac{\theta}{2} & \frac{1}{2}e^{-i\phi}\sin\theta \\ \frac{1}{2}e^{i\phi}\sin\theta & \sin^2\frac{\theta}{2} \end{pmatrix}. \tag{4.4}$$

Since any 2×2 complex Hermitian matrix can be expressed in terms of the identity matrix and the Pauli matrices, 2×2 mixed states—that is, 2×2 positive semidefinite matrices with trace 1—can be represented by the Bloch sphere. We write a Hermitian

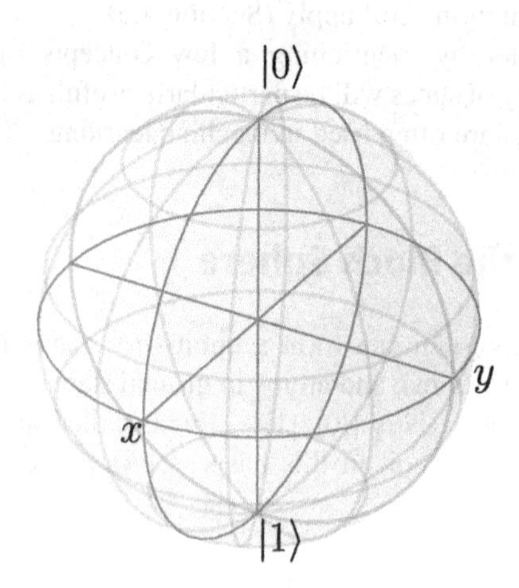

Figure 4.1 The Bloch sphere.

matrix as a real linear combination of $\{I, \sigma_x, \sigma_y, \sigma_z\}$, then we impose the positive semidefinite and trace 1 assumptions. Thus, a density matrix is written as $\rho = \frac{1}{2}(I + \mathbf{s}\sigma)$, where σ is a vector of the Pauli matrices, and $\mathbf{s}$ is called the Bloch vector. For pure states, this provides a one-to-one mapping to the surface of the Bloch sphere. Geometrically.

$$s = \begin{pmatrix} \sin\theta\cos\phi \\ \sin\theta\sin\phi \\ \cos\theta \end{pmatrix}, \tag{4.5}$$

where $\|\mathbf{s}\| = 1$. For mixed states, the Bloch vector lies in the interior of the Bloch ball.

With this interpretation, the computational basis corresponds to the Z axis of the Bloch sphere, and the eigenvectors of the Pauli matrix σ_z. We call the X axis of the Bloch sphere the diagonal basis, which corresponds to the eigenvectors of the Pauli matrix σ_x. The Y axis gives the circular basis, with the eigenvectors of the Pauli matrix σ_y. Pauli matrices are essentially rotations around the corresponding axes—for instance, about the X axis, we have $R_X = \mathrm{e}^{\mathrm{i}\theta\sigma_x/2}$.

Mixed states are in a way closer to classical states. The classical states are along the Z axis (the identity matrix is the origin). Quantum behavior emerges as we leave this axis. The two poles are of great interest: this is where pure states intersect with classical behavior.

The Bloch sphere stores an infinite amount of information, but neighboring points on the Bloch sphere cannot be distinguished reliably. Hence, we must construct states in a way to be able to tell them apart. This puts constraints on the storage capacity.

Multiple qubits are easy to construct as product spaces of individual qubits. The bases generalize from the one-qubit case. For instance, the computational basis of a two-qubit system is $\{|00\rangle, |01\rangle, |10\rangle, |11\rangle\}$. A generic state is in the superposition of these basis vectors:

$$|\psi\rangle = \alpha_{00}|00\rangle + \alpha_{01}|01\rangle + \alpha_{10}|10\rangle + \alpha_{11}|11\rangle, \tag{4.6}$$

where $|\alpha_{00}|^2 + |\alpha_{01}|^2 + |\alpha_{10}|^2 + |\alpha_{11}|^2 = 1$. Unfortunately, the geometric insights provided by the Bloch sphere for single qubits do not generalize to multiple qubits.

Entangled qubits may provide an orthonormal basis for the space of multiple qubits. For instance, the Bell states are maximally entangled, and they provide such a basis:

$$|\phi^+\rangle = \frac{1}{\sqrt{2}}(|00\rangle + |11\rangle), \tag{4.7}$$

$$|\phi^-\rangle = \frac{1}{\sqrt{2}}(|00\rangle - |11\rangle), \tag{4.8}$$

$$|\psi^+\rangle = \frac{1}{\sqrt{2}}(|01\rangle + |10\rangle), \tag{4.9}$$

$$|\psi^-\rangle = \frac{1}{\sqrt{2}}(|01\rangle - |10\rangle). \tag{4.10}$$

As is already apparent from the two-qubit case, the number of probability amplitudes is exponential in the number of qubits: an n-qubit system will have 2^n probability amplitudes. Thus, it is tempting to think that qubits encode exponentially more information than their classical counterparts. This is not the case: the measurement puts a limit on the maximum number of bits represented. This is known as Holevo's bound: n qubits encode at most n bits. Quantum information does not compress classical information.

4.2 Quantum Circuits

Quantum computing has several operational models, but only two are crucial for later chapters: quantum circuits and adiabatic quantum computing. Other models include topological quantum computing and one-way quantum computing. We begin with a discussion of quantum circuits.

Quantum circuits are the most straightforward analogue of classical computers: wires connect gates which manipulate qubits. The transformation made by the gates is always reversible: this is a remarkable departure from classical computers.

The circuit description uses simple diagrams to represent connections. Single wires stand for quantum states traveling between operations. Whether it is an actual wire, an optical channel or any other form of transmission is not important. Classical bits travel on double lines. Boxes represent gates and other unitary operations on one or multiple qubits. Since the transformation is always reversible, the number of input and output qubits is identical. Measurement is indicated by a box with a symbolic measurement device inside. Measurement in a quantum circuit is always understood to be a projective measurement, as ancilla systems can be introduced (Section 3.5). Measurement is performed in the computational basis, unless otherwise noted.

The most elementary single-qubit operation is a quantum NOT gate, which takes $|0\rangle$ to $|1\rangle$, and vice versa. A generic quantum state in the computational basis will change from $\alpha|0\rangle + \beta|1\rangle$ to $\alpha|1\rangle + \beta|0\rangle$.

Given a basis—usually the computational basis—we can represent quantum gates in a matrix form. If we denote the quantum NOT gate by the matrix X, it is defined as

$$X = \begin{pmatrix} 0 & 1 \\ 1 & 0 \end{pmatrix}. \tag{4.11}$$

Coincidentally, the quantum NOT gate is identical with the Pauli σ_x matrix (Equation 4.3).

Since states transform unitarily, such matrix representations for quantum gates are always unitary. The converse is also true: any unitary operation is a valid time evolution of the quantum state; hence, it defines a valid quantum gate. Whereas the only single-bit operation in classical systems is the NOT gate, single-qubit gates are in abundance. Not all of them are equally interesting, however.

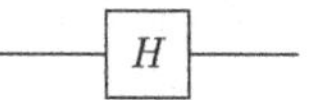

Figure 4.2 A Hadamard gate.

Two more gates, the Hadamard gate and the Z gate, are of special importance. The Hadamard gate is defined as

$$H = \frac{1}{\sqrt{2}}\begin{pmatrix} 1 & 1 \\ 1 & -1 \end{pmatrix}. \tag{4.12}$$

It is an idempotent operator: $H^2 = I$. It transforms elements of the computational basis to "halfway" between the elements. That is, it takes $|0\rangle$ into $(|0\rangle + |1\rangle)/\sqrt{2}$ and $|1\rangle$ into $(|0\rangle - |1\rangle)/\sqrt{2}$. In other words, sending through an element of the computational basis will put it in an equal superposition. Measuring the resulting state gives 0 or 1 with 50% probability. The symbol for a Hadamard gate in a circuit is show in Figure 4.2. Most gates follow this scheme: a box containing an acronym for the gate represents it in a quantum circuit.

The Z gate is defined as

$$Z = \begin{pmatrix} 1 & 0 \\ 0 & -1 \end{pmatrix}. \tag{4.13}$$

It leaves $|0\rangle$ invariant, and changes the sign of $|1\rangle$. This is essentially a phase shift.

While there are infinitely many single-qubit operations, they can be approximated to arbitrary precision with only a finite set of these gates.

Moving on to multiple-qubit operations, we discuss the controlled-NOT (CNOT) gate next. The CNOT gate takes two input qubits: one is called the control qubit, the other is the target qubit. Depending on the value of the control qubit, a NOT operation may be applied on the target qubit. The control qubit will remain unchanged irrespective of the value either qubit inputs.

The matrix representation of a CNOT gate is a 4×4 matrix to reflect the impact on both qubits. It is defined as

$$\text{CNOT} = \begin{pmatrix} 1 & 0 & 0 & 0 \\ 0 & 1 & 0 & 0 \\ 0 & 0 & 0 & 1 \\ 0 & 0 & 1 & 0 \end{pmatrix}. \tag{4.14}$$

The CNOT gate is a generalized XOR gate: its action on a bipartite state $|A, B\rangle$ is $|A, B \oplus A\rangle$, where $\oplus$ is addition modulo 2—that is, a XOR operation. The representation of a CNOT gate in a quantum circuit is shown in Figure 4.3.

The composite system also obeys unitary evolution; hence, every operation on multiple qubits is described by a unitary matrix. This unitary nature implies reversibility: it establishes a bijective mapping between input and output bits. Given the output and the operations, we can always recover the initial state. This is generally not true for classical gates. For instance, a classical XOR gate is a surjection: given an output,

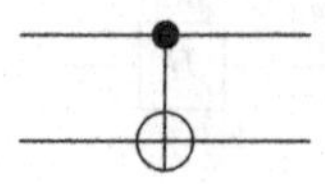

Figure 4.3 A CNOT gate.

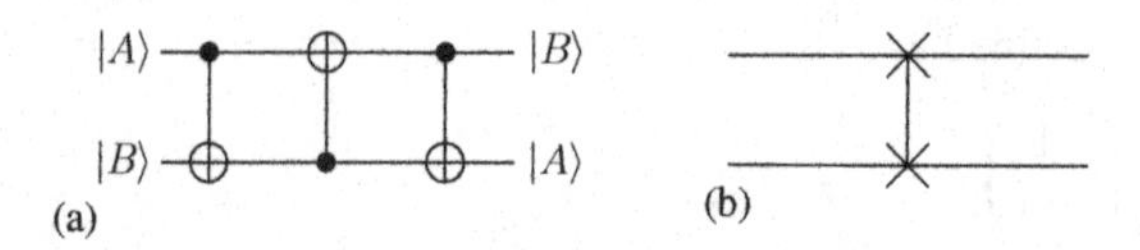

Figure 4.4 The quantum swap operation. (a) A sequence of three CNOT gates results in a swap operation. (b) The swap operation in a circuit.

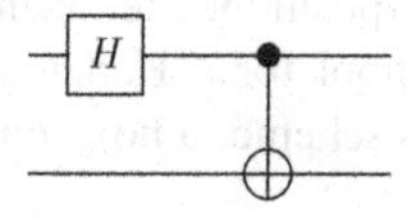

Figure 4.5 A circuit to generate Bell states from the computational basis.

there are two possible input configurations. The implications of this for learning algorithms are far-reaching, as they restrict the family of functions that can be learned directly, and leave nonlinearity or discontinuous manipulations to a measurement at the end of the unitary transformations. However, we must point out that arbitrary classical circuits can be made reversible by introducing ancilla variables.

If a logic gate is irreversible, information is erased—some of the information is lost as an operation is performed. In a reversible scenario, no information is lost. Landauer's principle establishes a relationship between the loss of information and energy: there is a minimum amount of energy required to change one bit of information, $kT \log 2$, where k is the Boltzmann constant, and T is the temperature of the environment. If a bit is lost, at least this much of energy is dissipated into the environment. Since quantum computers use reversible gates only, theoretically quantum computing could be performed without expending energy.

Three CNOT gates combine to create a swap operation (Figure 4.4). Given a state $|A, B\rangle$ in the computational basis, the gate operates as

$$|A, B\rangle \mapsto |A, A \oplus B\rangle \mapsto |A \oplus (A \oplus B), A \oplus B\rangle = |B, A \oplus B\rangle \tag{4.15}$$

$$\mapsto |B, (A \oplus B) \oplus B\rangle = |B, A\rangle. \tag{4.16}$$

A CNOT gate combined with a Hadamard gate generates the Bell states, starting with the computational basis. The computational basis is in a tensor product space, whereas the Bell states are maximally entangled: this gate combination allows us to create entangled states (Figure 4.5).

The Toffoli gate operates on three qubits, and it has special relevance to classical computations. It has the effect $|A, B, C\rangle \mapsto |A, B, C \oplus AB\rangle$—that is, it flips the third qubit if the first two control qubits are set to 1. In other words, it generalizes the

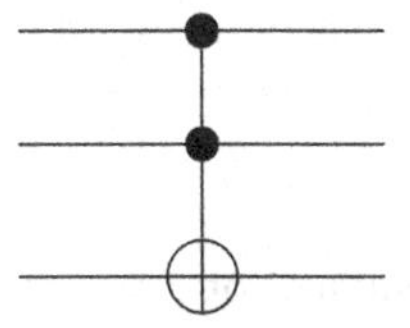

Figure 4.6 A Toffoli gate.

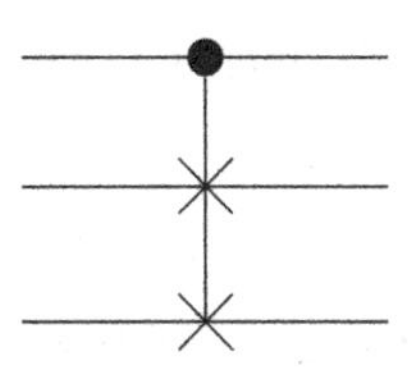

Figure 4.7 A Fredkin gate.

CNOT gate further, which itself is a generalized XOR operation. We denote the Toffoli gate by 2XOR. The circuit representation of the Toffoli gate is shown in Figure 4.6. Increasing the number of control qubits, we can obtain generic nXOR gates.

The Toffoli gate is an idempotent operator. The classical counterpart is universal: any classical circuit can be simulated by Toffoli gates: hence, every classical circuit has a reversible equivalent. The quantum Toffoli gate can simulate irreversible classical circuits; hence, quantum computers are able to perform any operation that classical computers can.

The next important gate, the Fredkin gate has three input qubits, but only one is for control (Figure 4.7). If the control bit is set to 1, the target qubits are swapped, otherwise the state is not modified. The Fredkin gate is idempotent. The control qubit in this case is an ancilla: its value is irrelevant to the computation performed. Its presence ensures that the gate is reversible.

The ancilla qubit may or may not change—it is not important for the result. Let us assume a function g represents the output in the ancilla; this is a garbage bit, its value is not important. By adding NOT gates, we can always ensure that the ancilla qubit starts from the state $|0\rangle$. The generic pattern of computation with an ancilla is thus

$$|x,0\rangle \mapsto |f(x),g(x)\rangle. \tag{4.17}$$

By appending a CNOT gate before the calculation, we make a copy of x that remains unchanged at the end of the transformation:

$$|x,0,0\rangle \mapsto |x,f(x),g(x)\rangle. \tag{4.18}$$

Let us add a fourth register y, and add the result with a CNOT operation:

$$|x,0,0,y\rangle \mapsto |x,f(x),g(x),y\oplus f(x)\rangle. \tag{4.19}$$

Apart from the final CNOT, the calculations did not affect y, and they were unitary; hence, if we apply the reverse operations, we get the following state:

$$|x, 0, 0, y \oplus f(x)\rangle. \tag{4.20}$$

This procedure is known as uncomputation—we uncompute the middle two registers that are used as scratch pads. We often omit the middle ancilla qubits, and simply write

$$|x, y\rangle \mapsto |x, y \oplus f(x)\rangle. \tag{4.21}$$

This formula indicates that reversible computing can be performed without production of garbage qubits.

4.3 Adiabatic Quantum Computing

In adiabatic quantum computing, the aim is to find the global minimum of a given function $f : \{0,1\}^n \mapsto (0, \infty)$, where $\min_x f(x) = f_0$ and $f(x) = f_0$ if and only if $x = x_0$—that is, there is a unique minimum (Kraus, 2013). We seek to find x_0. To do so, we consider the Hamiltonian

$$H_1 = \sum_{x \in \{0,1\}^n} f(x)|x\rangle\langle x|, \tag{4.22}$$

whose the unique ground state is $|x_0\rangle$. We take an initial Hamiltonian H_0 and the Hamiltonian in Equation 3.56 to adiabatically evolve the system. Thus, if we measure the system in the computational basis at the end of this process, we obtain x_0 (Farhi et al., 2000).

The gap between the ground state and the first excited states defines the computational complexity: smaller gaps result in longer computational times. The gap depends on the initial and the target Hamiltonian. Adiabatic quantum computing speeds up finding an optimum by about a quadratic factor over classical algorithms. It is harder to argue whether exponential speedups are feasible. Van Dam et al. (2001) have already defined lower bounds for optimization problems.

Adiabatic quantum computing is equivalent to the standard gate model of quantum computing (Aharonov et al., 2004), meaning that adiabatic quantum algorithms will run on any quantum computer (Kendon et al., 2010). Physical implementation is becoming feasible: adiabatic quantum computing has already demonstrated quantum annealing with over 100 qubits (Boixo et al., 2014), although the results are disputed (Rønnow et al., 2014). A further advantage of adiabatic quantum computing is that it is more robust against environmental noise and decoherence than other models of quantum computing (Amin et al., 2009; Childs et al., 2001).

When it comes to machine learning, the advantage of adiabatic quantum computing is that it bypasses programming paradigms: the simulated annealing of the Hamiltonian is equivalent to minimizing a function—a pattern frequently encountered in learning formulations (see Section 11.2 and Chapter 14).

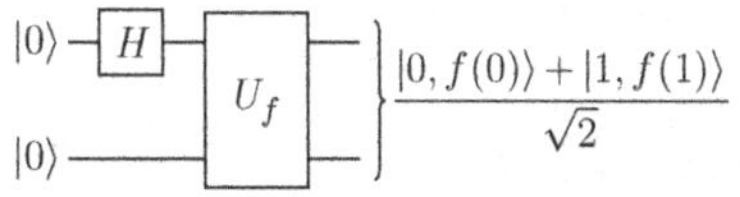

$$|0\rangle - H - U_f \quad \frac{|0,f(0)\rangle + |1,f(1)\rangle}{\sqrt{2}}$$
$$|0\rangle$$

Figure 4.8 A quantum circuit to demonstrate quantum parallelism. We evaluate the function $f : \{0,1\} \mapsto \{0,1\}$ with an appropriate unitary U_f. By putting the data register in the superposition of the computational basis, we evaluate the function in its entire domain in one step.

4.4 Quantum Parallelism

While it is true that n qubits represent at most n bits, there is a distinct advantage in using quantum circuits. Consider a function $f : \{0,1\} \mapsto \{0,1\}$. If an appropriate sequence of quantum gates is constructed, it is possible to transform an initial state $|x,y\rangle$ to $|x, y \oplus f(x)\rangle$ (see Equation 4.21). The first qubit is called the data register, and the second qubit is the target register. If $y = 0$, then we have $|x,f(x)\rangle$. We denote the unitary transformation that achieves this mapping by U_f.

Suppose we combine U_f with a Hadamard gate on the data register—that is, we calculate the function on $(|0\rangle + |1\rangle)/2$. If we perform U_f with $y = 0$, the resulting state will be

$$\frac{|0,f(0)\rangle + |1,f(1)\rangle}{\sqrt{2}}. \tag{4.23}$$

A single operation evaluated the function on both possible inputs (Figure 4.8). This phenomenon is known as quantum parallelism. Typically, if we measure the state, we have a 50% chance of obtaining $f(0)$ or $f(1)$—we cannot deduce both values from a single measurement. The fundamental question of quantum algorithms is how to exploit this parallelism without destroying the superposition.

This pattern generalizes to n qubits: apply a Hadamard gate on each data register, then evaluate a function. The method is also known as the Hadamard transform: it produces a superposition of 2^n states by using n gates. Adding a target register, we get the following state for an input state $|0\rangle^{\otimes n}|0\rangle$:

$$\frac{1}{\sqrt{2^n}} \sum_x |x\rangle|f(x)\rangle, \tag{4.24}$$

where the sum is over all possible values of x.

4.5 Grover's Algorithm

Grover's algorithm finds an element in an unordered set quadratically faster than the theoretical limit for classical algorithms. The sought element defines a function: the function is evaluated as true on an element if it is the sought element. Grover's algorithm uses internal calls to an oracle that tells the value of this function—that

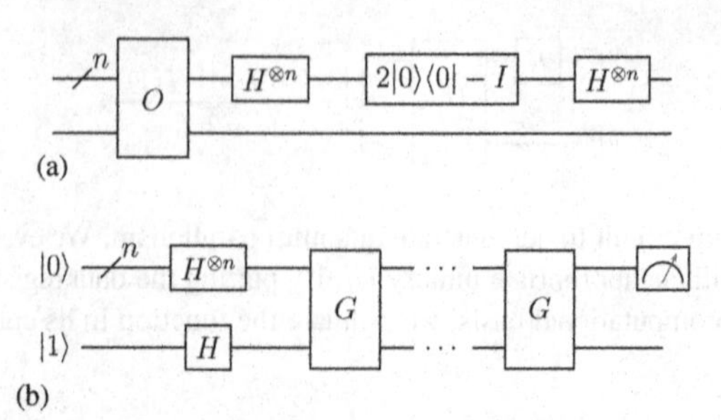

Figure 4.9 Circuit diagram of Grover's search. (a) The Grover operator. (b) The full circuit, where the Grover operator G is applied $O(\sqrt{N})$ times.

is, whether membership is true for a particular instance. The goal is then to use the smallest possible number of applications of the oracle to find all elements that test true (Figure 4.9).

Assume the data collection has N entries, with $n = \log N$ bits representing each entry. We start by applying a Hadamard transform on $|0\rangle^{\otimes n}$ to obtain the equal superposition state

$$|\psi\rangle = \frac{1}{\sqrt{n}} \sum_{x=0}^{n-1} |x\rangle. \tag{4.25}$$

We subsequently apply the Grover operator, also called the Grover diffusion operator, on this state a total of $O(\sqrt{N})$ times. We denote this operator as G, and it consists of four steps:

1. Apply a call to the oracle O.
2. Apply the Hadamard transform $H^{\otimes n}$.
3. Apply a conditional phase shift on the state with the exception of $|0\rangle$:

$$|x\rangle \mapsto -(-1)^{\delta_{x0}} |x\rangle. \tag{4.26}$$

4. Apply the Hadamard transform $H^{\otimes n}$ again.

The combined effect of steps 2-4 is

$$H^{\otimes n}(2|0\rangle\langle 0| - I)H^{\otimes n} = 2|\psi\rangle\langle\psi| - I, \tag{4.27}$$

where $|\psi\rangle$ is the equally weighted superposition of states in Equation 4.25. Thus the short way of writing the Grover operator is

$$G = (2|\psi\rangle\langle\psi| - I)O. \tag{4.28}$$

The implementation of the components of the Grover operator is efficient. The two Hadamard transforms require n operations each. The conditional phase shift is a controlled unitary operation and it requires $O(n)$ gates. The complexity of the oracle calls depends on the application, but only one call is necessary per iteration.

The key steps are outlined once more in Algorithm 1.

ALGORITHM 1 Grover's algorithm

```
Require: Initial state |0⟩^{⊗n}.
Ensure: Element requested.
  Initialize superposition by applying the Hadamard transform
  H^{⊗n}|0⟩^{⊗n}.
  for O(√N) times do
    Apply the Grover operator G
  end for
  Measure the system.
```

Adiabatic quantum computers are able to reproduce the algorithm in $O(\sqrt{N})$ time (Roland and Cerf, 2002; see Section 3.56 and Chapter 14).

Durr and Hoyer (1996) extended Grover's search to find the minimum (or maximum) in an unsorted array. The algorithm calls Grover's search to find the index of an item smaller than a particular threshold. This entry is chosen as the new threshold, and the procedure is repeated. The overall complexity is the same as for Grover's algorithm. Since the array is unordered, if we are able to define the search space as discrete, this method finds the global optimum with high probability, which is tremendously useful in machine learning algorithms.

4.6 Complexity Classes

Although some quantum algorithms are faster than their best known classical counterparts, quantum computers are limited by the same fundamental constraints. Any problem that is solved by a quantum computer can be solved by a classical computer, given enough resources. A Turing machine simulates any quantum computer, and the Church-Turing thesis binds quantum algorithms.

To gain insight into the actual power of quantum algorithms, let us take a look at complexity classes. Quantum computers run only probabilistic algorithms, and the class of problems that are solved efficiently by quantum computers is called bounded error, quantum, polynomial time (BQP). Bounded error means that the result of the computation will be correct with high probability.

BQP is the subset of PSPACE, the set of problems that can be solved by a Turing machine using a polynomial amount of space. BQP's relationship to other well-known classes is unknown. For instance, it is not known whether BQP is within NP, the set of problems for which a solution can be verified in polynomial time. NP-complete problems, a subset of NP with problems at least as hard as the ones in NP, are suspected of being disjoint from BQP.

It is conjectured that BQP is strictly larger than P, the problems for which a solution can be calculated in polynomial time using a deterministic Turing machine.

4.7 Quantum Information Theory

Quantum information theory generalizes classical information theory to quantum systems. Quantum information theory investigates the elementary processes of how quantum information is stored, manipulated, and transmitted. Its boundaries are vague—quantum computing might be included under this umbrella term. Quantum information theory is a rich field of inquiry, but we will restrict ourselves to a few fundamental concepts which will be relevant later.

A core concept of classical information theory is entropy, which has a generalization for the quantum case. Entropy is a measure to quantify the uncertainty involved in predicting the value of a random variable. Classical systems use the Shannon entropy, which is defined as

$$H(X) = -\sum_x \mathbf{P}(X = x) \log \mathbf{P}(X = x). \tag{4.29}$$

Its value is maximum for the uniform distribution. In other words, elements of a uniform distribution are the most unpredictable; the entropy for this case is $H(X) = \log n$, where the distribution has n values.

If we extend this measure to a quantum system, the von Neumann entropy of a state/reduced state described by a density matrix ρ is given by

$$S(\rho) = -\mathrm{tr}(\rho \log \rho). \tag{4.30}$$

The von Neumann entropy quantifies randomness—the randomness in the best possible measurement.

A function on a matrix is defined on its spectral decomposition. Taking the decomposition of $\rho = \sum_i \lambda_i |i\rangle\langle i|$, we get

$$S(\rho) = -\sum_k \left\langle k \left| \left(\sum_i \lambda_i \log \lambda_i |i\rangle\langle i| \right) \right| k \right\rangle = -\sum_n \lambda_n \log \lambda_n. \tag{4.31}$$

The density matrix of a pure state is a rank 1 projection: it is an idempotent operator. Hence, its von Neumann entropy is zero. Only mixed states have a von Neumann entropy larger than zero. The von Neumann entropy is invariant to basis change.

Bipartite pure states are entangled if and only if their reduced density matrix is mixed. Hence, applying the von Neumann entropy of a state/reduced state described a good quantifier of entanglement. Maximally entangled states were mentioned in Section 3.3. The formal definition is that a state is maximally entangled if its von Neumann entropy is $\log n$, where n is the dimension of the Hilbert state. Contrast this with the maximum entropy in the classical case.

Not all concepts in quantum information theory have classical counterparts. Quantum distinguishability is an example. In principle, telling two classical sequences of bits apart is not challenging. This is generally not true for quantum states, unless they are orthogonal. For instance, consider two states, $|0\rangle$ and $(|0\rangle + |1\rangle)/\sqrt{2}$. A single measurement will not tell which state we are dealing with. Measuring the second one will give 0 with probability $1/2$. Repeated measurements are necessary to identify the state.

Since the two states, ρ and σ, are, in general, not orthogonal, we cannot perform projective measurements, and we must restrict ourselves to a positive operator–valued measure (POVM). With this in mind, fidelity measures the distinguishability of two probability distributions—the probability distributions after measurement on the two states. The selection of POVMs matters; hence, we seek the best measurement that distinguishes the two states the most. For pure states, fidelity thus measures the overlap between two states, $|\psi\rangle$ and $|\phi\rangle$:

$$F(\psi, \phi) = |\langle\phi|\psi\rangle|. \tag{4.32}$$

This resembles the cosine similarity used extensively in machine learning (Section 2.2).

Fidelity on generic mixed or pure states is defined as

$$F(\rho, \sigma) = \operatorname{tr}\left(\sqrt{\sqrt{\rho}\sigma\sqrt{\rho}}\right). \tag{4.33}$$

The basic properties of fidelity are as follows:

1. Although it is not obvious from the definition, fidelity is symmetric in the arguments, but it does not define a metric.
2. Its range is in $[0, 1]$, and it equals 1 if and only if $\rho = \sigma$.
3. $F(\rho_1 \otimes \rho_2, \sigma_1 \otimes \sigma_2) = F(\rho_1, \sigma_1)F(\rho_2, \sigma_2)$.
4. It is invariant to unitary transformations: $F(U\rho U^\dagger, U\sigma U^\dagger) = F(\rho, \sigma)$.

Quantum state tomography is a concept related to quantum distinguishability where we are given an unknown state ρ, and we must identify what state it is. Quantum process tomography generalizes this to identifying an unknown unitary evolution; this procedure is analogous to learning an unknown function (Chapter 14).

Part Two

Classical Learning Algorithms

Unsupervised Learning

5

Unsupervised learning finds structures in the data. Labels for the data instances or other forms of guidance for training are not necessary. This makes unsupervised learning attractive in applications where data is cheap to obtain, but labels are either expensive or not available. Finding emergent topics in an evolving document collection is a good example: we do not know in advance what those topics might be. Detecting faults, anomalous instances in a long time series, is another example: we want to find out if something went wrong and if it did, then we would like to know when.

A learning algorithm, having no guidance, must identify structures on its own, relying solely on the data instances. Perhaps the most obvious approach is to start by studying the eigenvectors of the data, leading to geometric insights into the most prevalent directions in the feature space. Principal component analysis builds on this idea, and multidimensional scaling reduces the number of dimensions to two or three using the eigenstructure (Section 5.1). Dimensionality reduction is useful, as we are able to extract a visual overview of the data, but multidimensional scaling will fail to find nonlinear structures. Manifold learning extends the method to more generic geometric shapes in the high-dimensional feature space (Section 5.2).

If we study the distances between data instances, we are often able to find groups of similar ones. These groups are called clusters, and for data sets displaying simple geometric structures, K-means and hierarchical clustering methods detect them easily (Sections 5.3 and 5.4). If the geometry of the data instances is more complex, density-based clustering can help, which in turn is similar to manifold learning methods (Section 5.5).

Unsupervised learning is a vast field, and this chapter barely offers a glimpse. The algorithms in this chapter are the most relevant to quantum methods that have already been published.

5.1 Principal Component Analysis

Let us assume that the data matrix X, consisting of the data instances $\{\mathbf{x}_1, \ldots, \mathbf{x}_N\}$, has a zero mean. Principal component analysis looks at the eigenstructure of $X^\top X$. This $d \times d$ square matrix, where d is the dimensionality of the feature space, is known as the empirical sample covariance matrix in the statistical literature.

Quantum Machine Learning. http://dx.doi.org/10.1016/B978-0-12-800953-6.00001-1

If we calculate the eigendecomposition of $X^{\top}X$, we can arrange the normalized eigenvectors in a new matrix W. If we denote the diagonal matrix of eigenvalues by Λ, we have

$$X^{\top}X = W\Lambda W^{\top}. \tag{5.1}$$

We arrange the eigenvalues in Λ in decreasing order, and match the eigenvectors in W.

The principal component decomposition of the data matrix X is a projection to the basis given by the eigenvectors:

$$T = XW. \tag{5.2}$$

The coordinates in T are arranged so that the greatest variance of the data lies on the first coordinates, with the rest of the variances following in decreasing order on the subsequent coordinates (Jolliffe, 1989).

Using W, we can also perform projection to a lower-dimensional space, discarding some principal components. Let W_K denote the matrix with only K eigenvectors, corresponding to the K largest eigenvalues. The projection becomes

$$T_K = XW_K. \tag{5.3}$$

The matrix T_K has a feature space different from that of T, having only K columns. Among all rank K matrices, T_K is the best approximation to T for any unitarily invariant norm (Mirsky, 1960). Hence, this projection also minimizes the total squared error $\|T - T_K\|^2$. When projection to two or three dimensions is performed, this method is also known as multidimensional scaling (Cox and Cox, 1994).

Oddly enough, quantum states are able to reveal their own eigenstructure, which is the foundation of quantum principal component analysis (Section 10.3).

5.2 Manifold Embedding

As principal component analysis and multidimensional scaling are based on the eigendecomposition of the data matrix X, they can only deal with flat Euclidean structures, and they fail to discover the curved or nonlinear structures of the input data (Lin and Zha, 2008).

Manifold learning means a broad range of algorithms which assume that data in a high-dimensional space align to a manifold in a space of much lower dimensions. The method may also assume which manifold it is—this is typical of two-dimensional or three-dimensional embeddings which project the data onto a sphere or a torus (Ito et al., 2000; Onclinx et al., 2009). Methods that embed data points in a higher-dimensional space but still lower than the original, do not typically assume a specific manifold. Instead, they have assumptions about certain properties of the manifold—for instance, that it must be Riemannian (Lin and Zha, 2008).

As a prime example of manifold learning, Isomap extends multidimensional scaling to find a globally optimal solution for an underlying nonlinear manifold

(Tenenbaum et al., 2000). A shortcoming of Isomap is that it fails to find nonconvex embeddings (Weinberger et al., 2004). The procedure is outlined in Algorithm 2.

ALGORITHM 2 Isomap

```
Require: Initial data points {x_1, ..., x_N}.
Ensure: Low-dimensional embedding of data points.
  Find neighbors of each data point.
  Construct neighborhood graph.
  Estimate geodesic distances between all data points on the
  manifold by computing their shortest path distances in the graph.
  Minimize distance between geodesic distance and embedding.
  Euclidean distance matrices using eigendecomposition.
```

A major drawback of Isomap is its computational complexity. Finding the neighbors has $O(N^2)$ complexity, calculating the geodesic distances with Dijkstra's algorithm has $O(N^2 \log N)$ steps, and the final eigendecomposition is cubic.

If the size of the matrix that is subject to decomposition changes, updates to the embedding are necessary: iterative update is also possible by extensions (Law and Jain, 2006; Law et al., 2004). A new data point may introduce new shortest paths, and this is addressed by a modified version of Dijkstra's algorithm. Overall time complexity is the same as that of the original.

An extension called L-Isomap chooses landmark points to improve the run time (De Silva and Tenenbaum, 2003). Landmark selection avoids expensive, quadratic programming computations (Silva et al., 2006). Other extensions of Isomap are too numerous to mention, but this method highlights the crucial ideas in manifold learning methods that are useful for developing their quantum version (Section 10.4).

5.3 *K*-Means and *K*-Medians Clustering

The K-means algorithm—also called the K-nearest neighbors algorithm—is a method to cluster data instances on the basis of their pairwise distances into K partitions. It tries to minimize overall intracluster variance.

The algorithm starts by partitioning the input points into K initial sets, either at random or using some heuristic data. It then calculates the centroid of each set as follows:

$$\mathbf{c} = \frac{1}{N_c} \sum_{j=1}^{N_c} \mathbf{x}_j, \tag{5.4}$$

where N_c is the number of vectors in the subset. It constructs a new partition by associating each point with the closest centroid. The centroid-object distances are computed by the cosine dissimilarity or by other distance functions. Then the centroids are recalculated for the new clusters, and the process is repeated by alternate application of these two steps until their is convergence, which is obtained when the points no longer switch clusters. The convergence is toward a local minimum (Bradley and Fayyad, 1998).

The algorithm does not guarantee a global optimum for clustering: the quality of the final solution depends largely on the initial set of clusters. In text classification, however, where extremely sparse feature spaces are common, K-means has been proved to be highly efficient (Steinbach et al., 2000).

Unlike linear models, K-means does not divide the feature space linearly; hence, it tends to perform better on linearly inseparable problems (Steinbach et al., 2000). The most significant disadvantage is its inefficiency with regard to classification time. Linear classifiers consider a simple dot product, while in its simplest formulation, K-means requires the entire set of training instances ranked for similarity with the centroids (Lan et al., 2009).

Similarly to Equation 5.1 in principal component analysis, we may calculate the eigendecomposition of the covariance matrix of the data instances from $XX^\top$. The principal directions in this decomposition are identical to the cluster centroid subspace (Ding and He, 2004). This way, principal component analysis and K-means clustering are closely related, and by the appropriate projection, the search space reduces to where the global solution for clustering lies, helping to find near optimal solutions.

The quantum variant of K-means enables the fast calculation of distances, and an exponential speedup over the classical variant is feasible if we allow the input and output vectors to be quantum states (Section 10.5).

K-medians is a variation of K-means. In K-means, the centroid seldom coincides with a data instance: it lies between the data points. K-medians, on the other hand, calculates the median instead of the mean; hence, the representative element of a cluster is always a data instance. Unlike K-means, K-medians may not converge: it may oscillate between medians. K-medians is more robust and less sensitive to noise than K-means; moreover, it does not actually need the data points, it needs only the distances between them, and hence the algorithm works with using a Gram matrix alone (Section 7.4). Its computational complexity, however, is higher than for K-means (Park and Jun, 2009). It has, however, an efficient quantum version (Section 10.6).

5.4 Hierarchical Clustering

Hierarchical clustering is as simple as K-means, but instead of there being a fixed number of clusters, the number changes in every iteration. If the number increases, we talk about divisive clustering: all data instances start in one cluster, and splits are performed in each iteration, resulting in a hierarchy of clusters. Agglomerative clustering, on the other hand, is a bottom-up approach: each instance is a cluster at the beginning, and clusters are merged in every iteration. With use of either method, the hierarchy will have $N - 1$ levels (Hastie et al., 2008).

This way, hierarchical clustering does not provide a single clustering of the data, but provides clustering of $N - 1$ of them. It is up to the user to decide which one fits the purpose. Statistical heuristics are sometimes employed to aid the decision.

The arrangement after training, the hierarchy of clusters, is often plotted as a dendrogram. Nodes in the dendrogram represent clusters. The length of an edge between a cluster and its split is proportional to the dissimilarity between the split clusters. The popularity of hierarchical clustering is related to the dendrograms: these figures provide an easy-to-interpret view of the clustering structure.

Agglomerative clustering is more extensively researched than divisive clustering. Yet, the quantum variant (Section 10.7) is more apt for the divisive type. The classical divisive clustering algorithm begins by placing all data instances in a single cluster C_0. Then, it chooses the data instance whose average dissimilarity from all the other instances is the largest. This is the computationally most expensive step, having $\Omega(N^2)$ complexity in general. The selected data instance forms the first member of a second cluster C_1. Elements are reassigned from C_0 to C_1 as long as their average distance to C_0 is greater than that to C_1. This forms one iteration, after which we have two clusters, what remained from the original C_0, and the newly formed C_1. The procedure continues in subsequent iterations. The iterative splitting of clusters continues until all clusters contain only one data instance, or when it is no longer possible to transfer instances between clusters using the dissimilarity measure. Outliers are quickly isolated with this method, and unbalanced clusters do not pose a problem either.

5.5 Density-Based Clustering

Density-based clustering departs from the global objective functions in K-means and hierarchical clustering. Instead, this approach deals with local neighborhoods, resembling manifold learning methods. The data instances are assumed to alternate in high-density and low-density areas, the latter type separating the clusters. The shape of the clusters can thus be arbitrary, it does not even have to be convex (Kriegel et al., 2011).

Density-based spatial clustering of applications with noise is the most famous example (Ester et al., 1996). It takes two input parameters, ϵ and $N_{\min}$. The parameter ϵ defines a neighborhood $\{\mathbf{x}_j \in X | d(\mathbf{x}_i, \mathbf{x}_j) \leq \epsilon\}$ of the data instance $\mathbf{x}_i$. The minimum points parameter $N_{\min}$ defines a core object, a point with a neighborhood consisting of more elements than this parameter.

A point $\mathbf{x}_j$ is density-reachable from a core object $\mathbf{x}_i$ if a finite sequence of core objects between $\mathbf{x}_i$ and $\mathbf{x}_j$ exists such that each belongs to an ϵ- neighborhood of its predecessor. Two points are density-connected if both are density-reachable from a common core object.

Every point that is reachable from core objects can be factorized into maximally connected components serving as clusters. The points that are not connected to any core point can be considered as outliers, because they are not covered by any cluster. The run time complexity of this algorithm is $O(N \log N)$

With regard to the two parameters ϵ and N_{min}, there is no straightforward way to fit them to data. To overcome this obstacle, the algorithm orders the data instances to identify the clustering structure. This order augments the clusters with an additional data structure while remaining consistent with density-based clustering (Ankerst et al., 1999). Instead of just one point in the parameter space, this algorithm covers a spectrum of all different $\epsilon' \leq \epsilon$. The constructed ordering is used either automatically or interactively to find the optimal clustering.

Pattern Recognition and Neural Networks 6

A neural network is a network of units, some of which are designated as input and output units. These units are also called neurons. The units are connected by weighted edges—synapses. A unit receives a weighted input based on its connections, and generates a response of a univariate or multivariate form—this is the activation of a unit. In the simplest configuration, there is only one neuron, which is called a perceptron (Section 6.1).

If the network consists of more than one neuron, a signal spreads across the network starting from the input units, and either spreads one way toward the output units, or circulates in the network until it achieves a stable state. Learning consists of adjusting the weights of the connections.

Neural networks are inspired by the central nervous systems of animals, but they are not necessarily valid metaphors. Computational considerations make the update cycle different from that of natural systems. Furthermore, rigorous analysis from the theory of statistical learning also introduced changes that took neural networks further away from their biological counterparts.

Memorizing and recognizing a pattern are emergent processes of interconnected networks of simple units. This approach is called connectionist learning. The topology of the network is subject to infinite variations, which gave rise to hundreds of neural models (Sections 6.2 and 6.3).

While the diversity of neural networks is bewildering, a common characteristic is that the activation function is almost exclusively nonlinear in all neural network arrangements. The artificial neurons are essentially simple, distributed processing units. A clear delineation of subtasks is absent; each unit is performing a quintessentially similar computation in parallel. Recent advances in hardware favor this massively parallel nature of neural networks, allowing the development of extremely large connectionist models (Sections 6.4 and 6.5).

6.1 The Perceptron

The simplest type of neural network classifiers is the perceptron, consisting of a single artificial neuron (Rosenblatt, 1958). It is a linear discriminant: it cannot distinguish between linearly inseparable cases (Minsky and Papert, 1969).

Quantum Machine Learning. http://dx.doi.org/10.1016/B978-0-12-800953-6.00001-1

The single neuron in the perceptron works as a binary classifier (Figure 6.1). It maps an input $\mathbf{x} \in \mathbb{R}^d$ to a binary output value, which is produced by a Heaviside step function:

$$f(x) = \begin{cases} 1 & \text{if } \mathbf{w}^\top \mathbf{x} + b > 0, \\ 0 & \text{otherwise,} \end{cases} \tag{6.1}$$

where $\mathbf{w}$ is a vector weight—a weight in this vector corresponds to a feature in the feature space—and b is the bias term. The purpose of the bias term is to change the position of the decision plane. This function is called the activation function.

Training of the perceptron means obtaining the weight vector and the bias term to give the correct class for the training instances. Unfortunately, training is not guaranteed to converge: learning does not terminate if the training set is not linearly separable. The infamous XOR problem is an example: the perceptron will never learn to classify this case correctly (Figure 6.2). It is not surprising to find a set of four

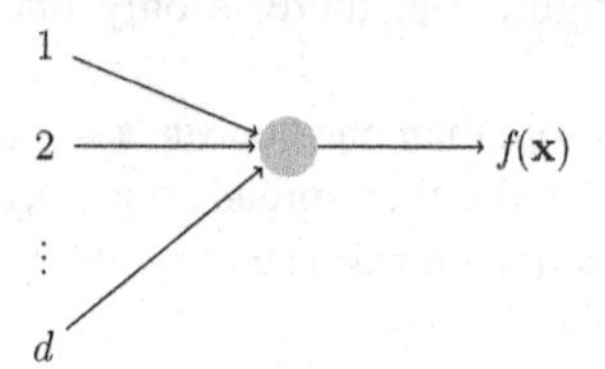

Figure 6.1 A perceptron input consists of the d-dimensional data instances. With a response function f, it produces an output $f(\mathbf{x})$

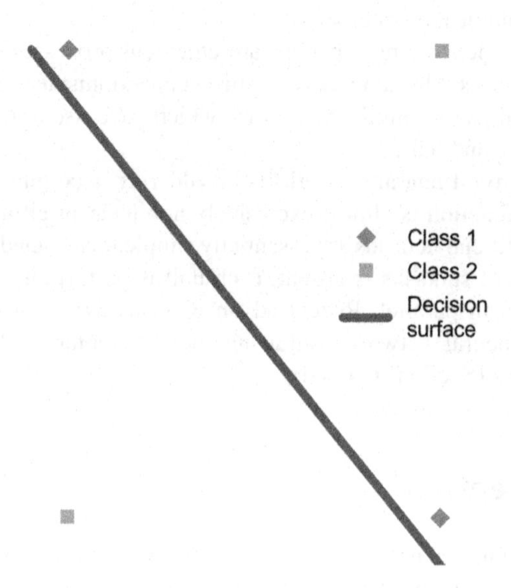

Figure 6.2 The XOR problem: points in a plan in this configuration can never be separated by a linear classifier.

points on the plane that the perceptron cannot learn, as the planar perceptron has a Vapnik-Chervonenkis dimension of three.

If the problem is linearly separable, however, the training of the perceptron will converge. The procedure is known as the delta rule, and it is a simple gradient descent. We minimize the error term

$$E = \frac{1}{2}\sum_{i=1}^{N}(y_i - f(\mathbf{x}_i))^2. \tag{6.2}$$

We initialize the weights and the threshold. Weights may be initialized to zero or to a random value. We take the simple partial derivative in w_j in the support of f:

$$\frac{\partial E}{\partial w_j} = -(y_i - f(\mathbf{x}_i))x_j. \tag{6.3}$$

The change to **w** should be proportional to this, yielding the updated formula for the weight vector:

$$\Delta w_j = \gamma (y_i - f(\mathbf{x}_i))\mathbf{x}_i, \tag{6.4}$$

where γ is a predefined learning rate.

The capacity of a linear threshold perceptron for large d is two bits per weight (Abu-Mostafa and St. Jacques, 1985; MacKay, 2005). This capacity is vastly expanded by the quantum perceptron (Section 11.2).

6.2 Hopfield Networks

A Hopfield network is a simple assembly of perceptrons that is able to overcome the XOR problem (Hopfield, 1982). The array of neurons is fully connected, although neurons do not have self-loops (Figure 6.3). This leads to $K(K-1)$ interconnections if there are K nodes, with a w_{ij} weight on each. In this arrangement, the neurons transmit signals back and forth to each other in a closed-feedback loop, eventually settling in stable states.

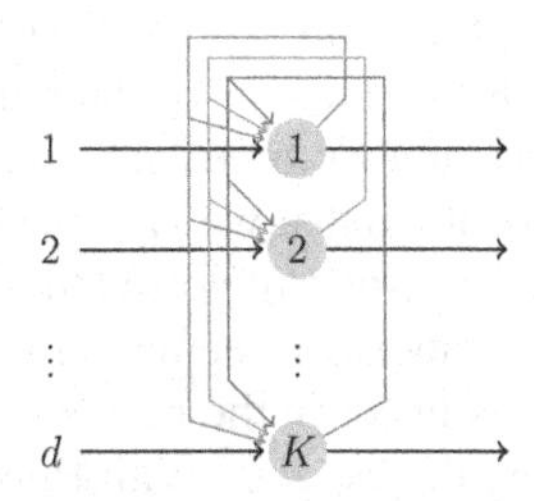

Figure 6.3 A Hopfield network with the number of nodes K matching the number of input features d.

An important assumption is that the weights are symmetric, $w_{ij} = w_{ji}$, for neural interactions. This is unrealistic for real neural systems, in which two neurons are unlikely to act on each other symmetrically.

The state s_i of a unit is either $+1$ or -1. It is activated by the following rule:

$$s_i = \begin{cases} +1 & \text{if } \sum_j w_{ij} s_j \geq \theta_i, \\ -1 & \text{otherwise,} \end{cases} \tag{6.5}$$

where θ_i is a threshold value corresponding to the node. This activation function mirrors that of the perceptron.

The activation of nodes happens either asynchronously or synchronously. The former case is closer to real biological systems: a node is picked to start the update, and consecutive nodes are activated in a predefined order. In synchronous mode, all units are updated at the same time, which is much easier to deal with computationally.

In a model called Hebbian learning, simultaneous activation of neurons leads to increments in synaptic strength between those neurons. The higher the value of a w_{ij} weight, the more likely that the two connected neurons will activate simultaneously. In Hopfield networks, Hebbian learning manifests itself in the following form:

$$w_{ij} = \frac{1}{N} \sum_{k=1}^{N} x_{ki} x_{kj}. \tag{6.6}$$

Here $\mathbf{x}_k$ is in binary representation—that is, the value $\mathbf{x}_{ki}$ is a bit for each i.

Hopfield networks have a scalar value associated with each neuron of the network that resembles the notion of energy. The sum of these individual scalars gives the "energy" of the network:

$$E = -\frac{1}{2} \sum_{i,j} w_{ij} s_i s_j + \sum_i \theta_i \, s_i. \tag{6.7}$$

If we update the network weights to learn a pattern, this value will either remain the same or decrease, hence justifying the name "energy." The quadratic interaction term also resembles the Hamiltonian of a spin glass or an Ising model, which some models of quantum computing can easily exploit (Section 14.3).

A Hopfield network is an associative memory, which is different from a pattern classifier, the task of a perceptron. Taking hand-written digit recognition as an example, we may have hundreds of examples of the number three written in various ways. Instead of classifying it as number three, an associative memory would recall a canonical pattern for the number three that we previously stored there. We may even consider an associative memory as a form of noise reduction.

The storage capacity of this associative memory—that is, the number of patterns that are stored in the network—is linear in the number of neurons. Estimates depend on the strategy used for updating the weights. With Hebbian learning, the estimate is about $N \leq 0.15K$. The quantum variant of Hopfield networks provides an exponential increase over this (Section 11.1).

6.3 Feedforward Networks

A Hopfield network is recurrent: the units form closed circles. If the number of output nodes required is lower, the storage capacity is massively improved by a nonrecurrent topology, the feedforward network. In a feedforward neural network, connections between the units do not form a directed cycle: information moves in only one direction, from the input nodes through an intermediate layer known as the hidden layer to the output nodes (Figure 6.4). Use of multiple hidden layers is sometimes advised. Rules for determining the number of hidden layers or the number nodes are either absent or ad hoc for a given application. The number of hidden units in the neural network affects the generalization performance, as the layers increase the model complexity (Rumelhart et al., 1994).

The input units represent the features in the feature space; hence, there are d nodes in this layer. The output units often represent the category or categories in a classification scenario; the number of nodes in this layer corresponds to this. If the output nodes are continuous and do not represent categories, we may view the network as a universal approximator for functions (Hornik et al., 1989). Following the model of Hebbian learning, the weights on the edges connecting the units between the layers represent dependency relations.

Feedforward networks use a variety of learning techniques. Often these are generalizations of the delta rule for training a perceptron—the most popular technique is called back-propagation.

The back-propagation algorithm starts with random weights w_{ij} on the synapses. We train the network in a supervised fashion, adjusting the weights with the arrival of each new training instance at the input layer. The changes in the weights are incremental and depend on the error produced in the output layer, where the output values are compared with the correct answer. Back-propagation refers to the feedback to the network through the adjustment of weights depending on the error

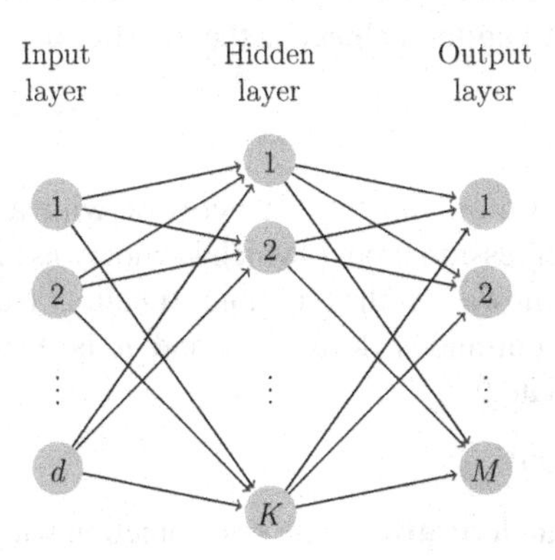

Figure 6.4 A feedforward neural network. The number of nodes in the input layers matches the dimensions of the feature space (d). The hidden layer has K nodes, and the output layer has M nodes, which matches the number of target classes.

function (Rumelhart et al., 1986). The scheme relies on a gradient descent; hence, the error function and the activation function should be differentiable.

This procedure repeats for a sufficiently large number of training cycles. In fact, one of the disadvantage of feedforward networks is that they need many training instances and also many iterations. The network weights usually stabilize in a local minimum of the error function, in which case we say that the network learned the target function. Given a sufficient number of hidden layers, the network is capable of learning linearly inseparable functions.

To describe the training procedure more formally, we start with loading the feature weights x_{ik} to the input units for classifying a test object $\mathbf{x}_i$. The activation of these units is then propagated forward through the network in subsequent layers. This activation function is usually nonlinear. In the final output layer, the units determine the categorization decision.

We study the squared error of the output node j for an input i:

$$E_{ij} = \frac{1}{2}(y_i - f_j(\mathbf{x}_i))^2, \tag{6.8}$$

where $f_j(\mathbf{x}_i)$ represents the response of the node j.

Let us consider a network with only one hidden layer of K nodes and with M output nodes. Let us split the weights w_{ij} into two subsets, α_{ik}, $0 \geq i \leq N$, $1 \geq k \leq K$, and β_{ij}, $0 \geq i \leq K$, $1 \geq j \leq M$. The weights α_{ij} are on the edges leading from the input layer to the hidden layer, and the weights β_{ij} are on the edges leading from the hidden layer to the output layer. Then, we can write the output of a node in the hidden layer for an input vector $\mathbf{x}_i$ as

$$z_k = g(\alpha_{0k} + \alpha_k^\top \mathbf{x}_i), \tag{6.9}$$

and the final output as

$$f_j(\mathbf{x}_i) = g(\beta_{0j} + \beta_j^\top \mathbf{z}), \tag{6.10}$$

where $\mathbf{z}$ is a vector of the z_k entries, and g denotes the activation function—a nonlinear, differentiable function. A common choice is the logistic function:

$$g(x) = \frac{1}{1 + e^{-x}}. \tag{6.11}$$

With this choice of activation function, a network with a single hidden layer is identical to the logistic regression model, which is widely used in statistical modeling. This function is also known as the sigmoid function (see the kernel types in Section 7.4). It has a continuous derivative, and it is also preferred because its derivative is easily calculated:

$$g'(x) = g(x)(1 - g(x)). \tag{6.12}$$

We calculate the partial derivative of the error function with respect to the weights:

$$\frac{\partial E_{ij}}{\beta_{jk}} = -(y_i - f_j(\mathbf{x}_i))g'(\beta_j^\top z_i)z_{ki}, \tag{6.13}$$

and

$$\frac{\partial E_{ij}}{\alpha_{kl}} = -\sum_{m=1}^{M}(y_i - f_m(\mathbf{x}_i))g'(\beta_m^\top z_i)\beta_{mk}g'(\alpha_k^\top \mathbf{x}_i)x_{il}. \quad (6.14)$$

Only the absolute value of the gradients matters. With a learning rate γ that decreases over the iterations, we have the updated formulas:

$$\beta_{jk} = \beta_{jk} - \gamma \sum_{i=1}^{N} \frac{\partial E_{ij}}{\beta_{jk}}, \quad (6.15)$$

$$\alpha_{kl} = \alpha_{kl} - \gamma \sum_{i=1}^{N} \frac{\partial E_{ij}}{\alpha_{kl}}. \quad (6.16)$$

Just as in the delta rule of the perceptron, in a learning iteration, we calculate the error term of the current weights: this is the forward pass in which the partial derivatives gain their value. In the backward pass, these errors are propagated to the weights to compute the updates of the weights, hence the name back-propagation.

As the objective function is based on the error terms in Equation 6.8, the surface on which we perform the gradient descent is nonconvex in general. Hence, it is possible that the procedure only reaches a local minimum; there are no guarantees that a global optimum will be reached.

The weights in the network are difficult to interpret, although in some cases rules can be extracted on the basis of their absolute value and sequence (Lu et al., 1996). Quantum neural networks replace weights with entanglement (Section 11.3).

The storage capacity for a network with a single hidden layer is still linear in the number of neurons. The coefficient, however, is better than in a Hopfield network, as N patterns need only N nodes in the hidden layer (Huang and Babri, 1998). A two-layer network can store $O(K^2/m)$ patterns, where m is the number of output nodes (Huang, 2003).

Back-propagation is a slow procedure; hence, developing faster training mechanisms is of great importance. One successful example is extreme learning machines, which perform a stochastic sampling of hidden nodes in the training iterations (Huang et al., 2006). Speed is also an advantage of quantum neural networks, with the added benefit that they may also find a global optimum with the nonconvex objective function.

6.4 Deep Learning

A long training time is a major issue that prevented the development of neural networks with several layers of hidden nodes for a long time. Deep learning overcomes this problem by computational tricks, including learning by layers and exploiting massively parallel architectures.

In deep learning, the representation of information is distributed: the layers in the network correspond to different levels of abstraction. Increasing the number of layers and the number of nodes in the layers leads to different levels of abstraction. This contrasts with shallow architectures, such as support vector machines, in which a single input layer is followed by a single layer of trainable coefficients. Deep learning architectures have the potential to generalize better with less human intervention, automatically building the layers of abstraction that are necessary for complex learning tasks (Bengio and LeCun, 2007).

The layers constitute a hierarchy: different concepts are learned from other concepts in a lower level of abstraction. Hence, toward the end of the learning pipeline, high-level concepts emerge. The associated loss function is often nonconvex (Bengio and LeCun, 2007).

Deep learning often combines unsupervised and supervised learning. Thus, deep learning algorithms make use of both unlabeled and labeled data instances, not unlike semisupervised learning. Unsupervised learning helps the training process by acting as a regularizer and aid to optimization (Erhan et al., 2010). Deep learning architectures are prone to overfitting because the additional layers allow the modeling of rare or even ad hoc dependencies in the training data. Attaching unsupervised learners to preprocessing or postprocessing which is unrelated to the main model helps overcome this problem (Hinton et al., 2012).

6.5 Computational Complexity

Given the diversity of neural networks, we cannot possibly state a general computational complexity for these learning algorithms. The complexity is topology-dependent, and it also depends on the training algorithm, and the nature of the neurons.

A Hopfield network's computational complexity will depend on the maximum absolute value of the weights, irrespective of whether the update function is synchronous or asynchronous (Orponen, 1994). The overall complexity is $O(N^2 \max_{i,j} |w_{ij})$.

In a feedforward network with N training instances and a total of w weights, each epoch requires $O(Nw)$ time. The number of epochs can be exponential in d, the number of input nodes (Han et al., 2012, p. 404).

If we look at computational time rather than time complexity, parallel computing can greatly decrease the amount of time that back-propagation takes to converge. To improve parallel efficiency, we must use a slight simplification: the update of weights in Equations 6.15 and 6.16 introduces dependencies that prevent parallelism. If we batch the updates to calculate the error over many training instances before the update is done, we remove this bottleneck.

In this batched algorithm, the training data are broken up into equally large parts for each thread in the parallel architecture. The threads perform the forward and backward passes, and corresponding weights are summed locally for each thread. At regular training intervals, the updates are shared between the threads to refresh the weights. The communication involved is limited to sharing the updates.

Simulated annealing helps speed up training, which also ensures convergence to the global optimum.

Gradient descent and back-propagation extend to training a deep learning network. As we noted earlier, their computational cost is high, but they are easy to implement and they are guaranteed to converge to a local minimum. The computational cost further increases with the sheer number of free parameters in a deep learning model: the number of abstraction layers, the number of nodes in the various layers, learning rates, initialization of weight vectors—these have to be tuned over subsequent independent training rounds. Massively parallel computer architectures enable the efficient training of these vast networks. There is little data dependency between individual neurons; hence, graphics processing units and similar massively parallel vector processors are able to accelerate the calculations, often reducing the computational time by a factor of 10-20 (Raina et al., 2009). Computations further scale out in a distributed network of computing nodes, enabling the training of even larger deep learning models.

Supervised Learning and Support Vector Machines

7

The primary focus in this chapter is on support vector machines, the main learning algorithm that derives from the Vapnik-Chervonenkis (VC) theory. Before discussing various facets of support vector machines, however, we take a brief detour to another, more simplistic supervised learner, the K-nearest neighbors (Section 7.1). This algorithm has an efficient quantum variant, and it also applies to a form of regression (Chapter 8), which in turn has an entirely different quantum formulation.

The development of support vector machines started with optimal margin hyperplanes that separate two classes with the highest expected generalization performance (Section 7.2). Soft margins allow noise training instances in cases where the two classes are not separable (Section 7.3).

With the use of using kernel functions, nonlinearity is addressed, allowing the embedding of data into a higher-dimensional space, where they become linearly separable, but still subject to soft margins (Section 7.4). "Feature space" in the machine learning literature refers to the high-dimensional space describing the characteristics of individual data instances (Section 2.2). In the context of support vector machines, however, this space is called *input space*. The reason is that the input space is often mapped to higher-dimensional space to tackle nonlinearities in the data, and the embedding space is called feature space. This is further explained in Section 7.4.

One quantum variant relies on a slightly altered formulation of support vector machines that uses least-squares optimization, which we discuss in Section 7.5.

Support vector machines achieve an outstanding generalization performance (Section 7.6), although the extension to multiclass problems is tedious (Section 7.7).

We normally choose loss functions to have a convex objective function, primarily for computational reasons (Section 7.8). A quantum formulation may liberate support vector machines from this constraint. Computational complexity can be estimated only in certain cases, but it is at least polynomial both in the number of training instances and in the number of dimensions (Section 7.9). Not surprisingly, the quantum variant will improve this.

Quantum Machine Learning. http://dx.doi.org/10.1016/B978-0-12-800953-6.00001-1

7.1 *K*-Nearest Neighbors

In the K-nearest neighbors algorithm, calculations only takes place when a new, unlabeled instance is presented to the learner. Seeing the new instance, the learner searches for the K most nearby data instances—this is the computationally expensive step that is easily accelerated by quantum methods (Section 12.1). Among those K instances, the algorithm will select the class which is the most frequent.

The construction of a K-nearest neighbor classifier involves determining a threshold K indicating how many top-ranked training objects have to be considered. Larkey and Croft (1996) used $K = 20$, while others found $30 \leq K \leq 45$ to be the most effective (Joachims, 1998; Yang and Chute, 1994; Yang and Liu, 1999).

The K-nearest neighbors algorithm does not build an explicit, declarative representation of the category c_i, but it has to rely on the category labels attached to the training objects similar to the test objects. The K-nearest neighbors algorithm makes a prediction based on the training patterns that are closest to the unlabeled example. For deciding whether $\mathbf{x}_j \in c_K$, the K-nearest neighbors algorithm looks at whether the K training objects most similar to $\mathbf{x}_j$ also are in c_K ; if the answer is positive for a large enough proportion of them, a positive decision is taken (Yang and Chute, 1994). This instance-based learning is a form of transduction.

7.2 Optimal Margin Classifiers

A support vector machine is a supervised learning algorithm which learns a given independent and identically distributed set of training instances $\{(\mathbf{x}_1, y_1), \ldots, (\mathbf{x}_N, y_N)\}$, where $y \in \{-1, 1\}$ are binary classes to which data points belong.

A hyperplane in $\mathbb{R}^d$ has the generic form

$$\mathbf{w}^\top \mathbf{x} - b = 0, \tag{7.1}$$

where $\mathbf{w}$ is the normal vector to the hyperplane, and the bias parameter b helps determine the offset of the hyperplane from the origin.

We assume that the data instances are linearly separable—that is, there exists a hyperplane that completely separates the data instances belonging to the two classes. In this case, we look for two hyperplanes such that there are no points in between and we maximize their distance. The area between the hyperplanes is the margin.

In its simplest, linear form, a support vector machine is a hyperplane that separates a set of positive examples from a set of negative examples with maximum margin. The distance between the two planes is $\|\frac{2}{\mathbf{w}}\|$; hence, minimizing $\mathbf{w}$ will lead to a maximal margin, which in turn leads to good generalization performance. The formula for the output of a linear support vector machine is

$$\hat{y}_i = \text{sign}(\mathbf{w}^\top \mathbf{x}_i + b), \tag{7.2}$$

where $\mathbf{x}_i$ is the ith training example. With this, the conditions for data instances for not falling into the margin are as follows:

$$\begin{aligned} \mathbf{w}^\top \mathbf{x}_i - b &\geq 1 \qquad \text{for } y_i = 1, \\ \mathbf{w}^\top \mathbf{x}_i - b &\leq -1 \qquad \text{for } y_i = -1. \end{aligned} \tag{7.3}$$

These conditions can be written briefly as

$$y_i(\mathbf{w}^\top \mathbf{x}_i - b) \geq 1, \quad i = 1, \ldots, N. \tag{7.4}$$

The optimization is subject to these constraints, and it seeks the optimal decision hyperplane with

$$\operatorname*{argmin}_{\mathbf{w},b} \frac{1}{2}\|\mathbf{w}\|^2. \tag{7.5}$$

The margin is also equal to the distance of the decision hyperplane to the nearest of the positive and negative examples. Support vectors are the training data that lie on the margin (Figure 7.1).

While this primal formulation of the linear case has a linear time complexity solution (Joachims, 2006), we more often refer to the dual formulation of the problem. To obtain the dual formulation, first we introduce Lagrange multipliers α_i to include the constraints in the objective function:

$$\operatorname*{argmin}_{\mathbf{w},b} \max_{\alpha_i \geq 0} \left(\frac{1}{2}\|\mathbf{w}\|^2 - \sum_{i=1}^{N} \alpha_i [y_i(\mathbf{w}^\top \mathbf{x}_i - b) - 1] \right). \tag{7.6}$$

The α_i corresponding to nonsupport vectors will be set to zero, as they do not make a difference in finding the saddle point of the expanded objective function. We denote

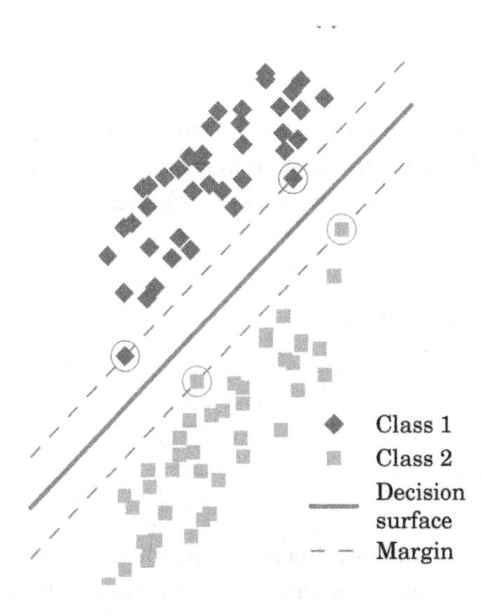

Figure 7.1 An optimal margin classifier maximizes the separation between two classes, where the separation is measured as the distance between the margins. Support vectors are the data instances that lie on the margins.

the objective function in Equation 7.6 by $L(\mathbf{w}, b, \boldsymbol{\alpha})$. Setting the derivatives of $\mathbf{w}$ and b to zero, we get

$$\begin{aligned} \frac{\partial L}{\partial w_i} &= w_i - \alpha_i y_i x_i = 0, \\ \frac{\partial L}{\partial b} &= \sum_{i=1}^{N} \alpha_i y_i = 0. \end{aligned} \tag{7.7}$$

From this, we find that the solution is a linear combination of the training vectors:

$$\mathbf{w} = \sum_{i=1}^{N} \alpha_i y_i \mathbf{x}_i. \tag{7.8}$$

Inserting this into the objective function in Equation 7.6, we are able to express the optimization problem solely in terms of α_i. We define the dual problem with the α_i multipliers,

$$\max_{\alpha_i} \sum_{i=1}^{N} \alpha_i - \frac{1}{2} \sum_{i,j} \alpha_i \alpha_j y_i y_j \mathbf{x}_i^\top \mathbf{x}_j \tag{7.9}$$

subject to (for any $i = 1, \ldots, N$)

$$\alpha_i \geq 0, \tag{7.10}$$

and from the minimization in b, the additional constraint is

$$\sum_{i=1}^{N} \alpha_i y_i = 0. \tag{7.11}$$

Eventually, only a few α_i will be nonzero. The corresponding $\mathbf{x}_i$ are the support vectors that lie on the margin. Thus, for the support vectors, $y_i(\mathbf{w}\mathbf{x}_i - b) = 1$. Rearranging the equation, we get the bias term $b = \mathbf{w}^\top \mathbf{x}_i - y_i$. We average it over all support vectors to get a more robust estimate:

$$b = \frac{1}{|\{i : \alpha_i \neq 0\}|} \sum_{i:\alpha_i \neq 0} (\mathbf{w}^\top \mathbf{x}_i - y_i). \tag{7.12}$$

7.3 Soft Margins

Not all classification problems are linearly separable. A few instances of one class may mingle with elements of the other classes. The hard-margin classifier introduced in Section 7.2 is not able to efficiently learn a model from such data. Even if the classes are separable, the dependence on the margin makes a hard-margin support vector machine sensitive to just a few points that lie near the boundary.

To deal with such cases, nonnegative slack variables help. These measure the degree to which a data instance deviates from the margin. The optimization becomes a tradeoff between maximizing the margin and controlling the slack variables. A parameter C balances the tradeoff—it is also called a penalty or cost parameter.

The condition for soft-margin classification with the slack variables becomes

$$\mathbf{w}^\top \mathbf{x}_i + b \geq 1 - \xi_i \quad \text{if } y_i = +1, \tag{7.13}$$

$$\mathbf{w}^\top \mathbf{x}_i + b \leq -1 + \xi_i \quad \text{if } y_i = -1. \tag{7.14}$$

The primal form of the optimization is to minimize $\|\mathbf{w}\|$ and the amount of deviation described by the slack variables, subject to the above constraints:

$$\min \frac{1}{2}\|\mathbf{w}\|^2 + C\sum_i \xi_i. \tag{7.15}$$

The corresponding Lagrangian, analogous to Equation 7.6, is

$$\underset{\mathbf{w},\boldsymbol{\xi},b}{\operatorname{argmin}} \max_{\boldsymbol{\alpha},\boldsymbol{\beta}} \left(\frac{1}{2}\|\mathbf{w}\|^2 + C\sum_{i=1}^N \xi_i - \sum_{i=1}^N \alpha_i [y_i(\mathbf{w}^\top \mathbf{x}_i - b) - 1 + \xi_i] - \sum_{i=1}^N \beta_i \xi_i \right). \tag{7.16}$$

with $\alpha_i, \beta_i \geq 0$. Taking the partial derivatives, we derive relations similar to Equation 7.7, with the additional equality $C - \beta_i - \alpha_i = 0$, which together with $\beta_i \geq 0$ implies that $\alpha_i \leq C$.

Thus, the dual formulation is similar to the hard-margin case:

$$\max_{\alpha_i} \sum_{i=1}^N \alpha_i - \frac{1}{2}\sum_{i,j} \alpha_i \alpha_j y_i y_j \mathbf{x}_i^\top \mathbf{x}_j \tag{7.17}$$

subject to

$$\begin{aligned} &0 \leq \alpha_i \leq C, \quad i = 1, \ldots, N, \\ &\sum_{i=1}^N \alpha_i y_i = 0. \end{aligned} \tag{7.18}$$

Nonzero slack variables occur when an α_i reaches its upper limit. The cost parameter C acts as a form of regularization: if the cost of misclassification is higher, a more accurate model is sought with increased complexity—that is, with a higher number of support vectors.

7.4 Nonlinearity and Kernel Functions

Maximum margin classifiers with soft margins were already an important step forward in machine learning, as they have an exceptional generalization performance (Section 7.6). What makes support vector machines even more powerful is that they are not restricted to linear decision surfaces.

The dual formulation enables a nonlinear kernel mapping that maps the data instances from an input space into a higher-dimensional—possibly even infinite-dimensional—embedding space, which is called feature space in this context. The core idea is to replace the inner product $\mathbf{x}_i^\top \mathbf{x}_j$ in the objective function in Equation 7.17 by a function that retains many properties of the inner product, yet which is nonlinear. This function is called a kernel.

We embed the data instances with an embedding function ϕ that classifies points as

$$y_i(\mathbf{w}^\top \phi(\mathbf{x}_i) + b) \geq 1 - \xi_i, \quad \xi_i \geq 0, \quad i = 1, \ldots, N. \tag{7.19}$$

Within these constraints, we look for the solution of the following optimization problem:

$$\min \frac{1}{2} \|w\|^2 + C \sum_{i=1}^{N} \xi_i. \tag{7.20}$$

The objective is identical to the soft-margin case (Equation 7.15).

As in the previous sections, we introduce Lagrangian multipliers to accommodate the constraints in the minimization problem. The partial derivatives in $\mathbf{w}$, b, and ξ define a saddle point of the Lagrangian, with which the dual formulation becomes the following quadratic programming problem:

$$\max_{\alpha_i} \sum_{i=1}^{N} \alpha_i - \frac{1}{2} \sum_{i,j} \alpha_i \alpha_j y_i y_j K(\mathbf{x}_i, \mathbf{x}_j) \tag{7.21}$$

subject to

$$\begin{aligned} &0 \leq \alpha_i \leq C, \quad i = 1, \ldots, N, \\ &\sum_{i=1}^{N} \alpha_i y_i = 0. \end{aligned} \tag{7.22}$$

The function $K(\mathbf{x}_i, \mathbf{x}_j) = \phi(\mathbf{x}_i)^\top \phi(\mathbf{x}_j)$ is the kernel function, the dot product of the embedding space.

In the dual formulation, the kernel function bypasses calculation of the embedding itself. We use the kernel function directly to find the optimum and subsequently to classify new instances—we do not need to define an actual embedding. This is exploited by many kernels. For the same reason, the embedding space can be of infinite dimensions. Bypassing the ϕ embedding is often called the kernel trick.

Technically, any continuous symmetric function $K(\mathbf{x}_i, \mathbf{x}_j) \in L_2 \otimes L_2$ may be used as an admissible kernel, as long as it satisfies a weak form of Mercer's condition (Smola et al., 1998):

$$\int\int K(\mathbf{x}_i, \mathbf{x}_j) g(\mathbf{x}_i) g(\mathbf{x}_j) \geq 0 \quad \text{for all } g \in L_2(\mathbf{R}^d). \tag{7.23}$$

That is, all positive semidefinite kernels are admissible. This condition ensures that the kernel "behaves" like an inner product.

We do not need the ϕ embedding to classify new data instances either. For an arbitrary new data point $\mathbf{x}$, the decision function for a binary decision problem becomes

$$f(\mathbf{x}) = \text{sign}\left(\sum_{i=1}^{N} \alpha_i y_i K(\mathbf{x}_i, \mathbf{x}) + b\right). \tag{7.24}$$

This formula reveals one more insight: the learned model is essentially instance-based. The support vectors are explicitly present through the kernel function. Since the parameter α_i is controlling the instances, and there is an additional bias term b, it is not a pure form of transduction, but it is closely related. Since classification is based on a few examples in a single step, support vector machines are considered shallow learners, as opposed to the multilayer architecture of deep learning algorithms (Section 6.4).

Some important kernels are listed in Table 7.1. The linear kernel does not have an embedding; this case is identical to the one described in Section 7.3.

The polynomial kernel has a bias parameter c and a degree parameter d. It is a good example to illustrate the kernel trick. For $d = 2$, the explicit embedding of the input space into the feature space is given by

$$\phi(\mathbf{x}) = (x_1^2, \ldots, x_d^2, \sqrt{2}x_1x_2, \ldots, \sqrt{2}x_1x_{d-1}, \sqrt{2}x_2x_3, \ldots, \sqrt{2}x_{d-1}x_d, \tag{7.25}$$

$$\sqrt{2c}x_1, \ldots, \sqrt{2c}x_d, c)$$

The kernel trick bypasses this embedding into a $\binom{d+2}{2} = \frac{d^2+3d+2}{2}$-dimensional space, and allows the use of just $d + 1$ multiplications.

The radial basis function kernel has no explicitly defined embedding, and it operates in an infinite-dimensional space. The parameter γ controls the radius of the basis function. If no prior knowledge is available to adjust its value, $\gamma = 1/2\sigma^2$ is a reasonable default value, where σ is the standard deviation of the data.

This approach of infinite-dimensional feature spaces gave rise to combining wavelets with support vector machines, proving that these kernels satisfy the Mercer condition while also addressing niche classification problems (Wittek and Tan, 2011; Zhang et al., 2004).

A sigmoid kernel is also frequently cited ($\tanh(\gamma(\mathbf{x}_i, \mathbf{x}_j) + c)$, where γ and c are parameters), but this kernel is not positive definite for all choices of the parameters, and for the valid range of parameters, it was shown to be equivalent with the radial basis function kernel (Lin and Lin, 2003). A support vector machine with a sigmoid kernel is also equivalent to a two-layer neural network with no hidden layers—a perceptron (Section 6.1).

Table 7.1 Common Kernels

	Linear	**Polynomial**	**Radial Basis Function**
$K(\mathbf{x}_i, \mathbf{x}_j)$	$\mathbf{x}_i^\top \mathbf{x}_j$	$(\mathbf{x}_i^\top \mathbf{x}_j + c)^d$	$\exp(-\gamma \lVert \mathbf{x}_i - \mathbf{x}_j \rVert^2)$

The objective function of the dual formulation in Equation 7.21 depends on the data point via the kernel function. If we evaluate the kernel function on all possible data pairs, we can define a matrix with elements $k_{ij} = K(\mathbf{x}_i, \mathbf{x}_j)$, $i, j = 1, \ldots, N$. We call this the kernel or Gram matrix. Calculating it efficiently is key to deriving efficient algorithms, as it encodes all information about the data that a kernel learning model can obtain.

The Gram matrix is symmetric by definition. As the kernel function must be positive definite, so must the Gram matrix. Naturally, the Gram matrix is invariant to rotations of the data points, as inner products are always rotation-invariant.

Let K_1 and K_2 be kernels over $X \times X$, $X \subset \mathbb{R}^d$, $a \in \mathbb{R}^+$, let$f$ be a real-valued function on X and in $L_2(X)$, let K_3 be a kernel over $Y \times Y$, $Y \subset \mathbb{R}^d$, $\theta : X \to Y$, and let B be a symmetric positive semidefinite $N \times N$ matrix. Furthermore, let $p(x)$ be a polynomial with positive coefficients over $\mathbb{R}$. For all $x, z \in X, x', z' \in Y$, the following functions are also kernels (Cristianini and Shawe-Taylor, 2000):

1. $K(\mathbf{x}, \mathbf{z}) := K_1(\mathbf{x}, \mathbf{z}) + K_2(\mathbf{x}, \mathbf{z})$.
2. $K(\mathbf{x}, \mathbf{z}) := aK_1(\mathbf{x}, \mathbf{z})$.
3. $K(\mathbf{x}, \mathbf{z}) := K_1(\mathbf{x}, \mathbf{z})K_2(\mathbf{x}, \mathbf{z})$.
4. $K(\mathbf{x}, \mathbf{z}) := f(\mathbf{x})f(\mathbf{z})$.
5. $K(\mathbf{x}, \mathbf{z}) := K_3(\theta(\mathbf{x}), \theta(\mathbf{z}))$.
6. $K(\mathbf{x}, \mathbf{z}) := \mathbf{x}^\top B\mathbf{z}$.
7. $K(\mathbf{x}, \mathbf{z}) := p(K_1(\mathbf{x}, \mathbf{z}))$.
8. $K(\mathbf{x}, \mathbf{z}) := \exp(K_1(\mathbf{x}, \mathbf{z}))$.

These properties show that we can easily create new kernels from existing ones using simple operations.

The properties apply to kernel functions, but we can apply further transformations on the Gram matrix itself, as long as it remains positive semidefinite. We center the data—that is, move the origin of the feature space to the center of mass of the data instances. The sum of the norms of the data instances in the feature space is the trace of the matrix, which is, in turn, the sum of its eigenvalues. Centering thus minimizes the sum of eigenvalues.

While the Gram matrix is likely to be full-rank matrix, a subspace projection can be beneficial through a low-rank approximation. As in methods based on rank reduction through singular value decomposition, low-rank approximations act as a form of denoising the data.

7.5 Least-Squares Formulation

Least-squares support vector machines modify the goal function of the primal problem by using the l_2 norm in the regularization term (Suykens and Vandewalle, 1999):

$$\text{Minimize} \quad \frac{1}{2}\mathbf{w}^\top \mathbf{w} + \frac{\gamma}{2}\sum_{i=1}^{M} e_i^2 \tag{7.26}$$

subject to the *equality* constraints

$$y_i(\mathbf{w}^\top \phi(\mathbf{x}_i) + b) = 1 - e_i, \quad i = 1, \ldots, N. \tag{7.27}$$

The parameter γ plays the same role as the cost parameter C. Seeking the saddle point of the corresponding Lagrangian, we obtain the following least-squares problem:

$$\begin{pmatrix} 0 & \mathbf{1}^\top \\ \mathbf{1} & K + \gamma^{-1} I \end{pmatrix} \begin{pmatrix} b \\ \alpha \end{pmatrix} = \begin{pmatrix} 0 \\ \mathbf{y} \end{pmatrix}, \tag{7.28}$$

where K is the kernel matrix, and $\mathbf{1}$ is a vector of 1's. The least-squares support vector machine trades off zero α_i for nonzero error terms e_i, leading to increased model complexity.

7.6 Generalization Performance

To guarantee good generalization performance, we expect a low VC dimension to give a tight bound on the expected error (Section 2.5). Oddly, support vector machines may have a high or even infinite VC dimension. Yet, we are able to establish limits on the generalization performance.

The VC dimension of the learner depends on the Hilbert space $\mathcal{H}$ to which the nonlinear embedding maps. If the penalty parameter C is allowed to take all values, the VC dimension of the support vector machine is $\dim(\mathcal{H}) + 1$ (Burges, 1998). For instance, for a degree-two polynomial kernel, it is $\frac{d^2+3d+2}{2}$, and for a radial basis function kernel, this dimension is infinite.

Despite this, we are able to set a bound on the generalization error (Shawe-Taylor and Cristianini, 2004). With probability $1 - \delta$, the bound is

$$\frac{1}{CN} - \frac{\sqrt{-W(\boldsymbol{\alpha}^*)}}{CN\gamma^*} + \frac{4}{N\gamma^*}\sqrt{\mathrm{tr}(K)} + 3\sqrt{\frac{\ln(2/\delta)}{2N}}, \tag{7.29}$$

where $\boldsymbol{\alpha}^*$ is the solution of the dual problem of the soft-margin formulation, $W(\boldsymbol{\alpha}^*) = -\frac{1}{2}\sum_{i,j} \alpha_i \alpha_j y_i y_j \mathbf{x}_i^\top \mathbf{x}_j$, and $\gamma^* = \left(\sum_{i=1}^N \alpha_i^*\right)^{-1/2}$. In accordance with the general principles of structural risk minimization, sparsity will improve this bound.

7.7 Multiclass Problems

The formulations of support vector machines considered so far was originally designed for binary classification. There are two basic approaches for multiclass classification problems. One involves constructing and combining several binary classifiers, and the other involves directly considering all data in one optimization formulation. The latter approach, the formulation to solve multiclass support vector machine problems in one step has variables proportional to the number of classes. Thus, for multiclass support vector machine methods, either several binary classifiers have to be constructed or a larger optimization problem is needed. In general, it is computationally more expensive to solve a multiclass problem than a binary problem with the same amount of data (Hsu and Lin, 2002).

The earliest implementation used for multiclass classification was the one-against-all method. It constructs M models, where M is the number of classes. The ith support vector machine is trained with all of the examples in the ith class with positive labels, and all other examples with negative labels. Thus, given training data $\{(\mathbf{x}_1, y_1), \ldots, (\mathbf{x}_N, y_N)\}$, the ith model solves the following problem:

$$\min_{\mathbf{w}^i, b^i, \xi^i} \frac{1}{2}(\mathbf{w}^i)^\top \mathbf{w}^i + C\sum_{j=1}^{N} \xi_j^i, \tag{7.30}$$

$$(\mathbf{w}^i)^\top \phi(\mathbf{x}_j) + b^i \geq 1 - \xi_j^i \quad \text{if } y_i = i, \tag{7.31}$$

$$(\mathbf{w}^i)^\top \phi(\mathbf{x}_j) + b^i \leq -1 + \xi_j^i \quad \text{if } y_i \neq i, \tag{7.32}$$

$$\xi_j^i \geq 0, \; j = 1, \ldots, N. \tag{7.33}$$

After the problem has been solved, there are M decision functions:

$$(\mathbf{w}^1)^\top \phi(\mathbf{x}) + b^1 \tag{7.34}$$

$$\vdots \tag{7.35}$$

$$(\mathbf{w}^M)^\top \phi(\mathbf{x}) + b^M. \tag{7.36}$$

An $\mathbf{x}$ is in the class which has the largest value of the decision function:

$$\text{class of } \mathbf{x} = \arg\max_{i=1,\ldots,M} (\mathbf{w}^i)^\top \phi(\mathbf{x}) + b^i. \tag{7.37}$$

Addressing the dual formulation results in M N-variable quadratic problems having to be solved.

Another major method is the one-against-one method. This method constructs $M(M-1)/2$ classifiers, where each one is trained on data from two classes. For training data from the ith and the jth classes, the following binary classification problem is solved:

$$\min_{\mathbf{w}^{ij}, b^{ij}, \xi^{ij}} \frac{1}{2}(\mathbf{w}^{ij})^\top \mathbf{w}^{ij} + C\sum_{t} \xi_t^{ij}, \tag{7.38}$$

$$(\mathbf{w}^{ij})^\top \phi(\mathbf{x}_t) + b^{ij} \geq 1 - \xi_t^{ij} \quad \text{if } y_t = i, \tag{7.39}$$

$$(\mathbf{w}^{ij})^\top \phi(\mathbf{x}_t) + b^{ij} \leq -1 + \xi_t^{ij} \quad \text{if } y_t = j, \tag{7.40}$$

$$\xi_t^{ij} \geq 0, \quad \forall t. \tag{7.41}$$

Thus, $M(M-1)/2$ classifiers are constructed.

The one-against-all and the one-against-one methods were shown to be superior to the formulation to solve multiclass support vector machine problems in one step, because the latter approach tends to overfit the training data (Hsu and Lin, 2002).

7.8 Loss Functions

Let us revisit the objective function in soft-margin classification in Equation 7.15. A more generic form is given by

$$\min \frac{1}{2}\|\mathbf{w}\|^2 + C\sum_i L(y_i, f(\mathbf{x}_i)), \tag{7.42}$$

where $L(y_i, f(\mathbf{x}_i))$ is a loss function. The optimal loss function is the 0-1 loss, which has the value 0 if $y_i = f(\mathbf{x}_i)$, and 1 otherwise. This is a nondifferentiable function which also leads to a nonconvex objective function. The most commonly used function instead is the hinge loss:

$$L(y_i, f(\mathbf{x}_i)) = \max(0, 1 - y_i f(\mathbf{x}_i)), \tag{7.43}$$

where $f(\mathbf{x}_i)$ is without the threshold—it is the raw output of the kernel expansion. The positive semidefinite nature of the kernel function implies the convexity of the optimization (Shawe-Taylor and Cristianini, 2004, p. 216). Positive semidefinite matrices form a cone, where a cone is a subspace closed under addition and multiplication by nonnegative scalars, which implies the convexity.

The number of support vectors increases linearly with the number of training examples when using a convex loss function (Steinwart, 2003). Given this theoretical result, the lack of sparsity prevents the use of convex support vector machines in large-scale problems. Convex loss functions are also sensitive to noise, especially label noise, and outliers. These are the core motivating factors to consider nonconvex formulations.

Depending on how the examples with an insufficient margin are penalized, it is easy to derive a nonconvex formulation, irrespective of the kernel used. When nonconvex loss functions are used, sparsity in the classifier improves dramatically (Collobert et al., 2006), making support vector machines far more practical for large data sets. The ramp loss function is the difference of two hinge losses, controlling the score window in which data instances become support vectors; this leads to improved sparsity (Ertekin et al., 2011). We can also directly approximate the optimal 0-1 loss function, instead of relying on surrogates like the hinge loss (Shalev-Shwartz et al., 2010). If we can estimate the proportion of noise in the labels, we can also derive a nonconvex loss function, and solve the primal form directly with quasi-Newton minimization (Stempfel and Ralaivola, 2009). Nonconvex support vector machines are an example of the rekindled interest in nonconvex optimization (Bengio and LeCun, 2007). We discuss further regularized objective functions and loss functions in Section 9.4.

7.9 Computational Complexity

A support vector with a linear kernel can be trained in linear time using the primal formulation (Joachims, 2006), but in general, limits are difficult to establish. The quadratic optimization problem is usually solved by sequential minimal optimization of the dual formulation, which chooses a subset of the training examples to work with, and subsequent iterations change the subset. This way, the learning procedure avoids costly numerical methods on the full quadratic optimization problem. While there are no strict limits on the speed of convergence for training, sequential minimal optimization scales between linear and quadratic time in the number of training instances on a range of test problems (Platt, 1999).

The calculations are dominated by the kernel evaluation. Given a linear or polynomial kernel, the calculation of entry in the kernel matrix takes $O(d)$ time; thus, calculating the whole kernel matrix has $O(N^2d)$ time complexity. Solving the quadratic dual problem or the least-squares formulation has $O(N^3)$ complexity. Combining the two steps, the classical support vector machine algorithm has at least $O(N^2(N + d))$ complexity. This complexity can be mitigated by using spatial support structures for the data (Yu et al., 2003).

The quantum formulation yields an exponential speedup in these two steps, leading to an overall complexity of $O(\log(Nd))$ (Section 12.3).

Regression Analysis

8

In regression, we seek to approximate a function given a finite sample of training instances $\{(\mathbf{x}_1, y_1), \ldots, (\mathbf{x}_N, y_N)\}$, resembling supervised classification. Unlike in classification, however, the range of y_i is not discrete: it can take any value in $\mathbb{R}$. The approximating function f is also called the regression function.

A natural way of evaluating the performance of an approximating function f is the residual sum of squares:

$$E = \sum_{i=1}^{N} (y_i - f(\mathbf{x}_i))^2. \tag{8.1}$$

Minimizing this value over arbitrary families of functions leads to infinitely many solutions. Moreover, the residual sum of squares being zero does not imply that the generalization performance will be good. We must restrict the eligible functions to a smaller set.

Most often, we seek a parametric function estimate—that is, we seek the optimum in a family of functions characterized by parameters. The optimization is performed on the parameter space. We further refine function classes to linear estimates (Section 8.1) and nonlinear estimates (Section 8.2). The optimization process may also be regularized to ensure better generalization performance. The corresponding quantum method is based on process tomography, which optimizes fit over certain families of unitary transformations (Chapter 13).

Nonparametric regression follows a different paradigm. It requires larger sample sizes, but the predictor does not take a predefined form (Section 8.3).

8.1 Linear Least Squares

Linear least squares is a parametric regression method. If the feature space is $\mathbb{R}^d$, we assume that the approximating function has the form

$$f(\mathbf{x}, \boldsymbol{\beta}) = \boldsymbol{\beta}^\top \mathbf{x}, \tag{8.2}$$

where $\boldsymbol{\beta}$ is a parameter vector. This problem is also known as general linear regression. We seek to minimize the squared residual

Quantum Machine Learning. http://dx.doi.org/10.1016/B978-0-12-800953-6.00001-1

$$E(\boldsymbol{\beta}) = \sum_{i=1}^{N} (y_i - f(\mathbf{x}_i, \boldsymbol{\beta}))^2. \tag{8.3}$$

If we arrange the data instances in a matrix, the same formula becomes

$$\begin{aligned} E(\boldsymbol{\beta}) &= \|\mathbf{y} - X\boldsymbol{\beta}\|^2 = (\mathbf{y} - X\boldsymbol{\beta})^\top (\mathbf{y} - X\boldsymbol{\beta}) \\ &= \mathbf{y}^\top \mathbf{y} - \boldsymbol{\beta}^\top X^\top \mathbf{y} - \mathbf{y}^\top X\boldsymbol{\beta} + \boldsymbol{\beta}^\top X^\top X\boldsymbol{\beta}, \end{aligned} \tag{8.4}$$

where element i of the column vector $\mathbf{y}$ is y_i.

Differentiating this with respect to $\boldsymbol{\beta}$, we obtain the normal equations, which are written as

$$X^\top X\boldsymbol{\beta} = X^\top \mathbf{y}. \tag{8.5}$$

The solution is obtained by matrix inversion:

$$\boldsymbol{\beta} = (X^\top X)^{-1} X^\top \mathbf{y}. \tag{8.6}$$

An underlying assumption is that the matrix X has full rank—that is, the training instances are linearly independent.

The method also applies to quadratic, cubic, quartic, and higher polynomials. For instance, if $x \in \mathbb{R}$, then the approximating function becomes

$$f(x, \boldsymbol{\beta}) = \beta_0 + \beta_1 x + \beta_2 x^2. \tag{8.7}$$

For higher-order polynomials, orthogonal polynomials provide better results.

We may also regularize the problem, either to increase sparsity or to ensure the smoothness of the solution. The minimization problem becomes

$$E(\boldsymbol{\beta}) = \|\mathbf{y} - X\boldsymbol{\beta}\|^2 + \|\Gamma X\|^2, \tag{8.8}$$

where Γ is an appropriately chosen matrix, the Tikhonov matrix.

8.2 Nonlinear Regression

In a nonlinear approximation, the combination of the model parameters and the dependency on independent variables is not linear. Unlike in linear regression, there is no generic closed-form expression for finding an optimal fit of parameters for a given family of functions.

Support vector machines extend to nonlinear regression problems—this method is called support vector regression (Drucker et al., 1997; Vapnik et al., 1997). Instead of a binary value of y_i, the labels take an arbitrary real value.

The approximating function is a linear combination of nonlinear basis functions, the kernel functions. This linear combination is parameterized by the number of support vector vectors: as in the case of classification problems, the model produced by support vector regression depends only on a subset of the training data—that is, the parameters are independent of the dimensions of the space.

The support vector formulation implies the cost function of the optimization ignores training data instances close to the model prediction—the optimization is regularized. The formulation most often uses a hinge loss function for regression.

8.3 Nonparametric Regression

Parametric regression is indirect: it estimates the parameters of the approximating function. Nonparametric regression, on the other hand, estimates the function directly. Basic assumptions apply to the function: it should be smooth and continuous (Härdle, 1990). Otherwise, nonparametric modeling accommodates a flexible form of the regression curve.

Smoothing is a form of nonparametric regression that estimates the influence of the data points in a neighborhood so that their values predict the value for nearby locations. The local averaging procedure is defined as

$$f(\mathbf{x}) = \frac{1}{N}\sum_{i=1}^{N} w_i(\mathbf{x})y_i, \tag{8.9}$$

where $w_i(x)$ is the weight function describing the influence of data point y_i at $\mathbf{x}$.

For instance, the extension of the K-nearest neighbors algorithm to regression problems will assign a zero weight for instances not in the K-nearest neighbors of a target $\mathbf{x}$, and all the others will have an equal weight N/K. The parameter K regulates the smoothness of the curve.

Kernel smoothing regulates smoothness through the support of a kernel function. One frequent approximation is the Nadaraya-Watson estimator:

$$f(\mathbf{x}) = \frac{\sum_{i=1}^{N} K_h(\mathbf{x} - \mathbf{x}_i)y_i}{\sum_{i=1}^{N} K_h(\mathbf{x} - \mathbf{x}_i)}, \tag{8.10}$$

where h is the support or bandwidth of the kernel.

Spline smoothing aims to produce a curve without much rapid local variation. It optimizes a penalized version of the squared residuals:

$$E = \sum_{i=1}^{N}(y_i - f(\mathbf{x}_i))^2 + \lambda \int_{\mathbf{x}_1}^{\mathbf{x}_d} f''(\mathbf{x})^2 \mathrm{d}\mathbf{x}. \tag{8.11}$$

8.4 Computational Complexity

In linear least-squares regression, solving the normal equation (Equation 8.5) is typically done via singular value decomposition. The overall cost of this is $O(dN^2)$ steps. Matrix inversion has an efficient quantum variant, especially if the input and output are also quantum states (Section 10.3). Bypassing inversion, quantum process tomography directly seeks a unitary transformation over certain groups, which is also similar to parametric regression (Chapter 13).

In nonparametric regression, we usually face polynomial complexity. Using the K-nearest neighbors algorithm, we find the complexity is about cubic in the number of data points. The most expensive step is finding the nearest neighbors, which is done efficiently by quantum methods (Section 12.1).

Boosting

9

Boosting is an ensemble learning method that builds multiple learning models to create a composite learner. Boosting differs from bagging mentioned in Section 2.6: it explicitly seeks models that complement one another, whereas bagging is agnostic to how well individual learners deal with the data compared with one another.

Boosting algorithms iteratively and sequentially train weak classifiers with respect to a distribution, and add them to a final strong classifier (Section 9.1). As they are added, they are weighted to reflect the accuracy of each learner.

One learner learns what the previous one could not: the relative weights of the data instances are adjusted in each iteration. Examples that are misclassified by the previous learner gain weight, and examples that are classified correctly lose weight. Thus, subsequent weak learners focus more on data instances that the previous weak learners misclassified.

Sequentiality also means that the learners cannot be trained in parallel, unlike in bagging and other ensemble methods. This puts boosting at a computational disadvantage.

The variation between boosting algorithms is how they weight training data instances and weak learners. Adaptive boosting (AdaBoost) does not need prior knowledge of the error rate of the weak learners; rather, it adapts to the accuracies observed (Section 9.2).

Generalizing AdaBoost, a family of convex objective functions define a range of similar boosting algorithms that overcome some of the shortcomings of the original method (Section 9.3). Yet, convex loss functions will always struggle with outliers, and this provides strong motivation to develop nonconvex objective functions. These, however, always require computational tricks to overcome the difficulty of finding the optimum (Section 9.4).

9.1 Weak Classifiers

If the base learners are simple, they are referred to as decision stumps. A decision stump classifies an instance on the basis of the value of just a single input feature (Iba and Langley, 1992). Decision stumps have low variance but high bias. Yet, such simple learners perform surprisingly well on common data sets, for instance, compared with full decision trees (Holte, 1993).

Quantum Machine Learning. http://dx.doi.org/10.1016/B978-0-12-800953-6.00001-1

A decision tree classifier is a tree in which internal nodes are labeled by features. Branches departing from them are labeled by tests on the weight that the feature has in the test object. Leaves are labeled by target categories (Mitchell, 1997). The classifier categorizes an object $\mathbf{x}_i$ by recursively testing for the weights that the features labeling the internal nodes have in vector $\mathbf{x}_i$, until a leaf node is reached. The label of this node is then assigned to $\mathbf{x}_i$. Experience shows that for trees, a depth between four and eight works best for boosting (Hastie et al., 2008, p. 363).

A method for learning a decision tree for category c_k consists of the following divide-and-conquer strategy (Lan et al., 2009):

- Check whether all the training examples have the same label;
- If they do not, choose a feature j, partition the training objects into classes of objects that have the same value for j, and place each such class in a separate subtree.

The above process is recursively repeated on the subtrees until each leaf of the tree generated contains training examples assigned to the same category c_k. The selection of the feature j on which to operate the partition is generally made according to an information gain (Cohen and Singer, 1996; Lewis and Ringuette, 1994) or entropy (Sebastiani, 2002) criterion. A fully grown tree is prone to overfitting, as some branches may be too specific for the training data. Therefore, decision tree methods normally include a method for growing the tree and also one for pruning it, thus removing the overly specific branches (Mitchell, 1997).

The complexity of individual learners is less important: the diversity of learners usually leads to better overall performance of the composite classifier (Kuncheva and Whitaker, 2003). The diversity measure can be as simple as the correlation between the outputs of the weak learners; there does not appear to be a significant difference in applying various diversity measures. Decision stumps and small decision trees are examples of simple weak learners that can add to the variety.

Formally, in boosting, we iteratively train K weak classifiers, $\{h_1, h_2, \ldots, h_K\}$. The combined classifier weights the vote of each weak classifier as a function of its accuracy. In an iteration, we either change the weight of one weak learner, this process is known as corrective boosting, or change the weight of all of them, leading to totally corrective boosting.

9.2 AdaBoost

AdaBoost adapts to the strengths and weaknesses of its weak classifiers by emphasizing training instances in subsequent classifiers that were misclassified by previous classifiers. For a generic outline, see Algorithm 3.

AdaBoost initially assigns a weight w_i to each training instance. The value of the weight is $1/N$, where N is the number of training instances.

In each iteration, a subset is sampled from the training set. Subsequent training sets may overlap, and the sampling is done with replacement. The selection probability of an instance equals its current weight. Then we train a classifier on the sample—the error is measured on the same set. The error of a weak classifier's error rate at an iteration t is given by

$$E_t = \sum_{i=1}^{N} w_i \mathbb{1}(y_i \neq h_t(\mathbf{x}_i)). \tag{9.1}$$

Weights of the data instances are adjusted after training the weak classifier. If an instance is incorrectly classified, its weight will increase. If it is correctly classified, its weight will decrease. Hence, data instances with a high weight indicate that the weak learners struggle to classify it correctly. Formally, the weight of a correctly classified instance is multiplied by $E_t/(1 - E_t)$. Once all weights have been updated, they are normalized to give a probability distribution.

Since the probability is higher that difficult cases make it to a sample, more classifiers are exposed to them, and the chances are better that one of the learners will classify them correctly. Hence, classifiers complement one another.

ALGORITHM 3 AdaBoost

```
Require: Training and validation data, number of training
  iterations T
Ensure: Strong classifier
  Initialize weight distribution N over training samples as uniform
  distribution ∀i : d(i) = 1/N.
  for t = 1 to t = T do
    From the family of weak classifiers ℋ, find the classifier
    h_t that maximizes the absolute value of the difference of
    the corresponding weighted error rate E_t and 0.5 with respect
    to the distribution d(s): h_t = argmax_{h_t∈ℋ} |0.5 − E_t|, where
    E_t = Σ_{i=1}^{N} d(i)I(y_i ≠ h_t(x_i)).
    if |0.5 − E_t| ≤ β, where β is a previously chosen threshold, then
      Stop.
    end if
    α_t ← E_t/(1 − E_t)
    for i = 1 to N do
      d(i) ← d(i) exp{α_t[2I(y_i ≠ h_t(x_i)) − 1]}
      Normalize d(i) ← d(i)/Σ_{j=1}^{N} d(j)
    end for
  end for
```

In the composite classifier, a weight is also assigned to each weak learner. The lower the error rate of a classifier, the higher its weight should be. The weight is usually chosen as

$$\alpha_t = \frac{1}{2}\ln\frac{1 - E_t}{E_t}. \tag{9.2}$$

The error of the combined strong classifier on the training data approaches zero exponentially fast in the number of iterations if the weak classifiers do at least slightly better than random guessing. The error rate of the strong classifier will improve if any of the weak classifiers improve. This is in stark contrast with earlier boosting, bagging, and other ensemble methods.

If the Vapnik-Chervonenkis (VC) dimension of the weak classifiers is $d \geq 2$, then, after T iterations, the VC dimension of the strong classifier will be at most

$$2(d+1)(T+1)\log_2(e(T+1)). \tag{9.3}$$

Given the VC dimension, we may derive a formal upper limit on the generalization error with Vapnik's theorem, and hence establish the optimal number of iterations.

9.3 A Family of Convex Boosters

AdaBoost is a minimization of a convex loss function over a convex set of functions. The loss being minimized is exponential:

$$L(y_i, f(\mathbf{x}_i)) = \mathrm{e}^{-y_i f(\mathbf{x}_i)}, \tag{9.4}$$

and we are seeking a function

$$f(\mathbf{x}) = \sum_t w_t h_t(\mathbf{x}). \tag{9.5}$$

Finding the optimum is equivalent to a coordinate-wise gradient descent through a greedy iterative algorithm: it chooses the direction of steepest descent at a given step (Friedman et al., 2000).

Following this thought, we can, in fact, define a family of boosters by changing the loss function in Equation 9.4 (Duffy and Helmbold, 2000).

A convex loss will incur some penalty to points of small positive margin—that is, points that are correctly classified and close to the boundary. This is critical to obtaining a classifier of maximal margin, which ensures good generalization (Section 2.4). A good convex loss function will assign zero penalty to points of large positive margin, which are points correctly classified and that are also far from the boundary. A convex loss function applies a large penalty to points with large negative margin—that is, points which are incorrectly classified and are far from the boundary (Figure 9.1).

Thus, it is easy to see that any convex potential—a nonincreasing function in C^1 with a limit zero at infinity—is bound to suffer from classification noise (Long and Servedio, 2010). A label that is incorrect in the training data will incur a large negative margin, and it will pull the decision surface from away the optimum, leading to a distorted classifier.

Unbounded growth of negative margins is the key problem, explaining why the performance of AdaBoost is poor in the presence of label noise: its exponential label noise is oversensitive to mislabeling (Dietterich, 2000).

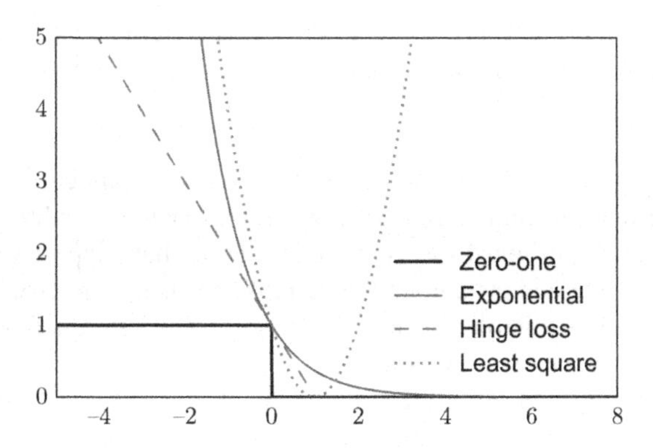

Figure 9.1 Convex loss functions compared with the optimal 0-1 loss.

The convex nature of the cost function enables the optimization to be viewed as a gradient descent to a global optimum (Mason et al., 1999). For instance, the sigmoid loss will lead to a global optimum:

$$L(y_i, f(\mathbf{x}_i)) = \frac{1}{N}\left[1 - \tanh(\lambda y_i f(\mathbf{x}_i))\right]. \tag{9.6}$$

Higher values of λ yield a better approximation of the 0-1 loss. Similarly, LogitBoost replaces the loss function with one that grows linearly with the negative margin, improving sensitivity (Friedman et al., 2000):

$$L(y_i, f(\mathbf{x}_i)) = \log\left(1 + e^{-y_i f(x_i)}\right). \tag{9.7}$$

While changing the exponential loss to an alternative convex function will improve generalization performance, a gradient descent will only give a linear combination of weak learners. Sparsity is not considered. To overcome this problem, we may regularize boosting similarly to limit the number of steps in the iterations (Zhang and Yu, 2005) or use a soft-margin like in support vector machines (Rätsch et al., 2001).

A simple form of regularization is an early stop—that is, limiting the number of iterations the boosting algorithm can take. It is difficult to find an analytical limit when the training should stop. Yet, if the iterations continue to infinity, AdaBoost will overfit the data, including mislabeled examples and outliers.

Another simple form of regularization is by shrinkage, in which the contribution of each weak learner is scaled by a learning rate parameter $0 < \nu < 1$. Shrinkage provides superior results to restricting the number of iterations (Copas, 1983; Friedman, 2001). Smaller values of ν result in a larger training risk in the same number of iterations. At the price of a larger number of iterations, a small learning rate leads to better generalization error.

To deal with noisy labels, we can introduce soft margins as in the case of support vector machines (Rätsch et al., 2001). Without such regularization, boosting is sensitive to even a low number of incorrectly labeled training examples (Dietterich, 2000). If we use L_2 regularization, we maintain the convexity of the objective function:

$$L(\mathbf{w}) + \lambda\|\mathbf{w}\|^2, \tag{9.8}$$

where $L(\mathbf{w}) = \sum_{i=1}^{N} L(y_i, f(\mathbf{x}_i))$, and dependence on $\mathbf{w}$ is implicit through f. This is not unlike the primary formulation of support vector machines with L_2 regularization of soft margins in the primal form. The difference is that support vector machines consider quadratic interactions among the variables. This picture changes in QBoost, the boosting algorithm suitable for quantum hardware, where quadratic interactions are of crucial importance (Section 14.2).

In LPBoost (Demiriz et al., 2002), the labels produced by the weak hypotheses create a new feature space. This way, we are no longer constrained by sequential training. All weights are modified in an iteration; this method is also known as totally corrective boosting. The objective function is regularized by a soft margin criterion. The model is more sensitive to the quality of weak learners, but the final strong classifier will be sparser. Overall performance and computational cost are comparable to those of AdaBoost.

The ideal nonconvex loss function, the 0-1 loss (Section 7.8), remains robust even if the labels are flipped in up to 50% of the training instances, provided that the classes are separable (Manwani and Sastry, 2013). None of the convex loss functions described above handle mislabeled instances well. Hence, irrespective of the choice of convex loss, the limits established by Long and Servedio (2010) apply, which is a strong incentive to look at nonconvex loss functions.

9.4 Nonconvex Loss Functions

SavageBoost bypasses the problem of optimizing the objective function containing the nonconvex loss (Masnadi-Shirazi and Vasconcelos, 2008). Instead, it finds the minimum conditional risk, which remains convex—the convexity of the loss function is irrelevant to this optimization. The conditional risk is defined as

$$C_L(\eta, f) = \eta L(f) + (1-\eta)L(-f). \tag{9.9}$$

The minimum conditional risk

$$C_L^\star(\eta) = \inf_f C_L(\eta, f) = C_\phi(\eta, f_L^\star) \tag{9.10}$$

must satisfy two properties. First, it must be a concave function of $\eta \in [0,1]$. Then, if $f_L^\star$ is differentiable, $C_L^\star(\eta)$ is also differentiable and, for any pair $(v, \hat{\eta})$ such that $v = f_L^\star(\hat{\eta})$,

$$C_L(\eta, v) - C_L^\star(\eta) = B_{-C_L^\star}(\eta, \hat{\eta}), \tag{9.11}$$

where

$$B_F(\eta, \hat{\eta}) = F(\eta) - F(\hat{\eta}) - (\eta - \hat{\eta})F'(\hat{\eta}) \tag{9.12}$$

is the Bregman divergence of the convex function F. This second condition provides insight into learning algorithms as methods for the estimation of the class posterior probability $\eta(\mathbf{x})$. The optimal function $f(\mathbf{x})$ that minimizes the conditional risk in Equation 9.9 is equivalent to a search for the probability estimate $\hat{\eta}(x)$ which minimizes the Bergman divergence of $-C_L^\star$ in Equation 9.11.

Minimizing the Bergman divergence imposes no restrictions on the convexity of the loss function. Masnadi-Shirazi and Vasconcelos (2008) derived a nonconvex loss function, the Savage loss:

$$L(y_i, f(\mathbf{x}_i)) = \frac{1}{(1 + \mathrm{e}^{2y_i f(\mathbf{x}_i)})^2}. \tag{9.13}$$

The Savage loss function quickly becomes constant as $m \to -\infty$, making it more robust to outliers and mislabeled instances. The corresponding SavageBoost algorithm is based on a gradient descent, and it is not totally corrective. SavageBoost converges faster than convex algorithms on select data sets.

TangentBoost for computer vision uses a nonconvex loss function (Masnadi-Shirazi et al., 2010):

$$L(y_i, f(\mathbf{x}_i)) = [2\arctan(y_i f(\mathbf{x}_i)) - 1]^2. \tag{9.14}$$

It was designed to retain the desirable properties of a convex potential loss: it is margin-enforcing with small penalty for correctly classified instances close to the decision boundary. Additionally, it has a bounded penalty for large negative margins. In this sense, it resembles Savage loss. Yet, it also penalizes points with large positive margin. This property might be useful in improving the margin, as points further from the boundary still influence the decision surface. In selected computer vision tasks, tangent loss performed marginally better than Savage loss. Since tangent loss is nonconvex, an approximate gradient descent estimates the optimum using the Gauss algorithm. Some nonconvex loss functions are shown in Figure 9.2.

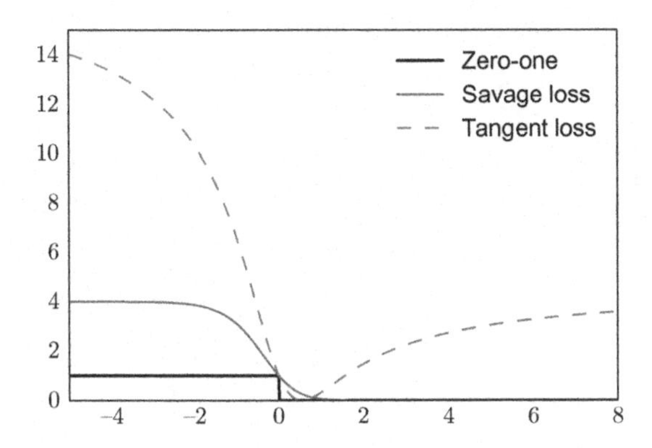

Figure 9.2 Nonconvex loss functions compared with the optimal 0-1 loss.

Part Three

Quantum Computing and Machine Learning

Clustering Structure and Quantum Computing

10

Quantum random access memory (QRAM) allows the storing of quantum states, and it can be queried with an address in superposition (Section 10.1). Calculating dot products and a kernel matrix relies on this structure (Section 10.2). These two components are not only relevant in unsupervised learning, they are also central to quantum support vector machines with exponential speedup (Section 12.3).

Quantum principal component analysis is the first algorithm we discuss to use QRAM: it retrieves quantum states which perform a self-analysis: an eigendecomposition of their own structure (Section 10.3). Quantum manifold learning takes this a step further, by initializing a data structure with geodesic distances (Section 10.4).

The quantum K-means algorithm may rely on classical or quantum input states; in the latter case, it relies on a QRAM and the quantum method for calculating dot products (Section 10.5). Quantum K-medians emerges from different ideas and it relies on Grover's search to deliver a speedup (Section 10.6). Quantum hierarchical clustering resembles this second approach (Section 10.7).

We summarize overall computational complexity for the various approaches in the last section of this chapter (Section 10.8).

10.1 Quantum Random Access Memory

A random access memory allows memory cells to be addressed in a classical computer: it is an array in which each cell of the array has a unique numerical address. A QRAM serves a similar purpose (Giovannetti et al., 2008).

A random access memory has an input register to address the cell in the array, and an output register to return the stored information. In a QRAM, the address and output registers are composed of qubits. The address register contains a superposition of addresses $\sum_j p_j|j\rangle_a$, and the output register will contain a superposition of information, correlated with the address register: $\sum_j p_j|j\rangle_a|D_j\rangle_d$.

Using a "bucket-brigade" architecture, a QRAM reduces the complexity of retrieving an item to $O(\log 2^n)$ switches, where n is the number of qubits in the address register.

The core idea of the architecture is to have qutrits instead of qubits allocated in each node of a bifurcation graph (Figure 10.1). A qutrit is a three-level quantum system. Let us label the three levels $|\text{wait}\rangle$, $|\text{left}\rangle$, and $|\text{right}\rangle$. During each memory call, each qutrit is in the $|\text{wait}\rangle$ state.

Quantum Machine Learning. http://dx.doi.org/10.1016/B978-0-12-800953-6.00001-1

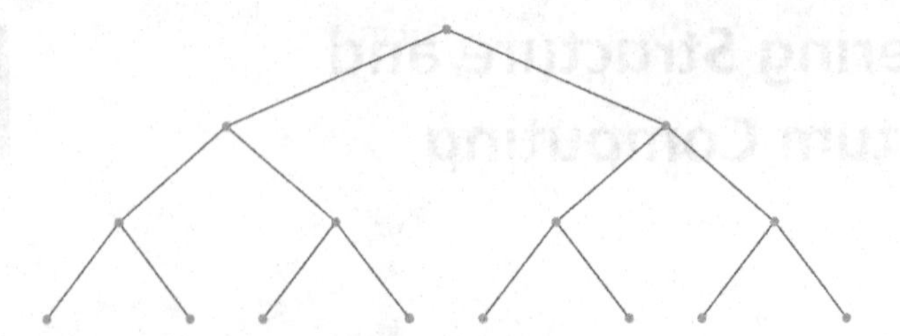

Figure 10.1 A bifurcation graph for QRAM: the nodes are qutrits.

The qubits of the address register are sent through the graph one by one. The $|\text{wait}\rangle$ state is transformed into $|\text{left}\rangle$ and $|\text{right}\rangle$, depending on the current qubit. If the state is not in $|\text{wait}\rangle$, it routes the current qubit. The result is a superposition of routes. Once the routes have thus been carved out, a bus qubit is sent through to interact with the memory cells at the end of the routes. Then, it is sent back to write the result to the output register. Finally, a reverse evolution on the states is performed to reset all of them to $|\text{wait}\rangle$.

The advantage of the bucket-brigade approach is the low number of qutrits involved in the retrieval: in each route of the final superposition, only $\log N$ qutrits are not in the $|\text{wait}\rangle$ state. The average fidelity of the final state if all qutrits are involved in the superposition is $O(1-\epsilon\log N)$ (Giovannetti et al., 2008).

10.2 Calculating Dot Products

In the quantum algorithm to calculate dot products, the training instances are presented as quantum states $|\mathbf{x}_i\rangle$. We do not require the training instances to be normalized, but the normalization must be given separately. To reconstruct a state from the QRAM, we need to query the memory $O(\log N)$ times.

To evaluate the dot product of two training instances, we need to do the following (Lloyd et al., 2013a):

- Generate two states, $|\psi\rangle$ and $|\phi\rangle$, with an ancilla variable;
- Estimate the parameter $Z = \|\mathbf{x}_i\|^2 + \|\mathbf{x}_j\|^2$, the sum of the squared norms of the two instances;
- Perform a projective measurement on the ancilla alone, comparing the two states.

Z times the probability of the success of the measurement yields the square of the Euclidean distance between the two training instances: $\|\mathbf{x}_i - \mathbf{x}_j\|^2$. We calculate the dot product in the linear kernel as $\mathbf{x}_i^\top \mathbf{x}_j = \frac{Z-\|\mathbf{x}_i-\mathbf{x}_j\|^2}{2}$.

The state $|\psi\rangle = \frac{1}{\sqrt{2}}(|0\rangle|\mathbf{x}_i\rangle + |1\rangle|\mathbf{x}_j\rangle)$ is easy to construct by querying the QRAM. We estimate the other state

$$|\phi\rangle = \frac{1}{Z}(\|\mathbf{x}_i\||0\rangle - \|\mathbf{x}_j\||1\rangle), \tag{10.1}$$

and the parameter Z together. We evolve the state

$$\frac{1}{\sqrt{2}}(|0\rangle - |1\rangle) \otimes |0\rangle \tag{10.2}$$

with the Hamiltonian

$$H = \left(\|\mathbf{x}_i\| |0\rangle\langle 0| + \|\mathbf{x}_j\| |1\rangle\langle 1| \right) \otimes \sigma_x. \tag{10.3}$$

The resulting state is

$$\begin{aligned} &\frac{1}{\sqrt{2}}\big[\cos(\|\mathbf{x}_i\|t)|0\rangle - \cos(\|\mathbf{x}_j\|t)|1\rangle \big] \otimes |0\rangle \\ &-\frac{\mathrm{i}}{\sqrt{2}}\big[\sin(\|\mathbf{x}_i\|t)|0\rangle - \sin(\|\mathbf{x}_j\|t)|1\rangle \big] \otimes |1\rangle. \end{aligned} \tag{10.4}$$

By appropriate choice of t ($\|\mathbf{x}_i\|t$, $\|\mathbf{x}_j\|t \ll 1$), and by measuring the ancilla bit, we get the state ϕ with probability $\frac{1}{2}Z^2t^2$, which in turns allows the estimation of Z. If the desired accuracy of the estimation is ϵ, then the complexity of constructing $|\phi\rangle$ and Z is $O(\epsilon^{-1})$.

If we have $|\psi\rangle$ and $|\phi\rangle$, we perform a swap test on the ancilla alone. A swap test is a sequence of a Hadamard gate, a Fredkin gate and another Hadamard gate which checks the equivalence of two states using an ancilla state (Buhrman et al., 2001). The circuit is shown in Figure 10.2.

The first transformation swaps $|\psi\rangle$ and $|\phi\rangle$ if the ancilla bit is $|1\rangle$. The Hadamard transformation is a one-qubit rotation mapping the qubit states $|0\rangle$ and $|1\rangle$ to two superposition states. More formally, the overall state after the first Hadamard transformation is

$$|a\rangle|\psi\rangle|\phi\rangle = \frac{1}{\sqrt{2}}(|0\rangle + |1\rangle)|\psi\rangle|\phi\rangle = \frac{1}{\sqrt{2}}(|0\rangle|\psi\rangle|\phi\rangle + |1\rangle|\psi\rangle|\phi\rangle). \tag{10.5}$$

We apply the Fredkin gate:

$$\frac{1}{\sqrt{2}}(|0\rangle|\psi\rangle|\phi\rangle + |1\rangle|\phi\rangle|\psi\rangle). \tag{10.6}$$

Then, we apply the Hadamard gate on the first qubit:

$$\frac{1}{2}\left[(|0\rangle + |1\rangle)\,|\psi\rangle|\phi\rangle + (|0\rangle - |1\rangle)\,|\phi\rangle|\psi\rangle\right] \tag{10.7}$$

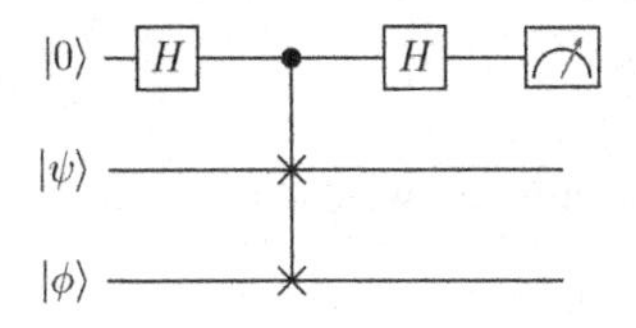

Figure 10.2 Quantum circuit of a swap test.

We rearrange the components for $|0\rangle$ and $|1\rangle$:

$$\frac{1}{2}\left[|0\rangle\left(|\psi\rangle|\phi\rangle+|\phi\rangle|\psi\rangle\right)+|1\rangle\left(|\psi\rangle|\phi\rangle-|\phi\rangle|\psi\rangle\right)\right]. \tag{10.8}$$

With this algebraic manipulation, we conclude that if $|\psi\rangle$ and $|\phi\rangle$ are equal, then the measurement in the end will always give us zero.

With the QRAM accesses and the estimations, the overall complexity of evaluating a single dot product $\mathbf{x}_i^\top \mathbf{x}_j$ is $O(\epsilon^{-1}\log N)$. Calculating the kernel matrix is straightforward. As the inner product in this formulation derives from the Euclidean distance, the kernel matrix is also easily calculated with this distance function.

Further generalization to nonlinear metrics approximates the distance function by qth-order polynomials, where we are given q copies of $|\mathbf{x}_i\rangle$ and $\mathbf{x}_j\rangle$ (Harrow et al., 2009; Lloyd et al., 2013a). Using quantum counting for the q copies, we evaluate $\left(\langle\mathbf{x}_i|\langle\mathbf{x}_j|\right)^{\otimes k} L(|\mathbf{x}_i\rangle|\mathbf{x}_j\rangle)^{\otimes k}$ for an arbitrary Hermitian operator L to get the approximation. Measuring the expectation value of L to accuracy ϵ has $O(\epsilon^{-1} q\log N)$ complexity.

A similar scheme uses pretty good measurements, which is just a form of positive operator–valued measures, to estimate the inner product and the kernel matrix (Gambs, 2008). This variant requires $\Theta(\epsilon^{-1}N)$ copies of each state.

10.3 Quantum Principal Component Analysis

Principal component analysis relies on the eigendecomposition of a matrix, which, in a quantum context, translates to simulating a Hamiltonian. The problem of simulating a Hamiltonian is as follows. Given the Hamiltonian of the form $H=\sum_{j=1}^{m} H_j$, the task is to simulate the evolution $\mathrm{e}^{\mathrm{i}Ht}$ by a sequence of exponentials $\mathrm{e}^{-\mathrm{i}H_j t'}$, requiring that the maximum error in the final state does not exceed some $\epsilon > 0$. We want to determine an upper bound on the number of exponentials required in the sequence (Berry et al., 2007).

The idea of approximating by a sequence of exponentials is based on the Baker-Campbell-Hausdorff formula, which states that

$$\mathrm{e}^Z = \mathrm{e}^X \mathrm{e}^Y, \tag{10.9}$$

where $Z=\log(\mathrm{e}^X\mathrm{e}^Y)$, and X and Y are noncommuting variables. A consequence of this is the Lie product formula

$$\mathrm{e}^{X+Y} = \lim_{n\to\infty} (\mathrm{e}^{X/n}\mathrm{e}^{Y/n})^n. \tag{10.10}$$

Thus, to simulate the Hamiltonian, we can calculate

$$\mathrm{e}^{\mathrm{i}Ht} \approx (\mathrm{e}^{\mathrm{i}H_1 t/n}\cdots\mathrm{e}^{\mathrm{i}H_m t/n})^n. \tag{10.11}$$

If the component Hamiltonians H_i act only locally on a subset of variables, calculating $\mathrm{e}^{\mathrm{i}H_i t}$ is more efficient. If the dimension of the Hilbert space that H_i acts on is d_i, then the number of operations to simulate $e^{iH_i t}$ is approximately d_i^2 (Lloyd, 1996). If the desired accuracy is ϵ, it will be bounded by

$$\epsilon \approx n \left(\sum_{j=1}^{m} d_i^2 \right) \leq nmd^2, \tag{10.12}$$

where $d = \max_i d_i$. This is the error term for the component Hamiltonians, and adjustment of $\epsilon/(nmd^2)$ is necessary to have an overall error of ϵ. Thus, if the Hamiltonians are local, to simulate the system, the time complexity is linear.

Let us consider arbitrary, but sparse Hamiltonians—that is, systems where the interactions of the components may not be local. A Hamiltonian H acting on n qubits is sparse if it has at most a constant number of nonzero entries in each row or column, and, furthermore, H is bounded by a constant. In this case, we can select an arbitrary positive integer k such that the simulation of the Hamiltonian requires $O((\log^* n)t^{1+1/2k})$ accesses to the matrix entries of H (Berry et al., 2007). Here $\log^* n$ is the iterated logarithm:

$$\log^* n = \begin{cases} 0 & \text{if } n \leq 1, \\ 1 + \log^*(\log n) & \text{if } n > 1, \end{cases} \tag{10.13}$$

that is, the number of times the logarithm function must be applied recursively before the result is less than or equal to 1. While the complexity is close to linear, sublinear scaling of the sparse simulation is not feasible (Berry et al., 2007).

To simulate nonsparse, arbitrary d-dimensional Hamiltonians, the typical method is the higher-order Trotter-Suzuki expansion, which requires $O(d \log d)$ time for a component Hamiltonian (Wiebe et al., 2010). This can be reduced to just $O(\log d)$ by applying a density matrix ρ as a Hamiltonian on another density matrix σ (Lloyd et al., 2013b). Using the partial trace over the first variable and the swap operator S, we get

$$\begin{aligned} \mathrm{tr_P}(e^{-iS\Delta t} \rho \otimes \sigma e^{iS\Delta t}) &= (\cos^2 \Delta t)\sigma + (\sin^2 \Delta t)\rho - i \sin \Delta t[\rho, \sigma] \\ &= \sigma - i\Delta t[\rho, \sigma] + O(\Delta t^2). \end{aligned} \tag{10.14}$$

Since the swap operator S is sparse, it can be performed efficiently. Provided we have n examples of ρ, we repeat swap operations on $\rho \otimes \sigma$ to construct the unitary operator $e^{-i\rho\Delta t}$, and thus we simulate the unitary time evolution $e^{-i\rho\Delta t}\sigma e^{i\rho\Delta t}$. To simulate $e^{-i\rho t}$ to accuracy ϵ, we need $O(t^2\epsilon^{-1}\|\rho - \sigma\|^2) \leq O(t^2\epsilon^{-1})$ steps, where $t = n\Delta t$ and $\|.\|$ is the supremum norm.

To retrieve ρ, we use a QRAM with $O(\log d)$ operations. Using $O(t^2\epsilon^{-1})$ copies of ρ, we are able to implement $e^{i\rho t}$ to accuracy ϵ in time $O(t^2/\epsilon \log d)$.

As the last step, we perform a quantum phase estimation algorithm using $e^{i\rho t}$. For varying times t, we apply this operator on ρ itself, which results in the state

$$\sum_i r_i |\chi_i\rangle\langle\chi_i| \otimes |\hat{r}_i\rangle\langle\hat{r}_i|, \tag{10.15}$$

where $|\chi_i\rangle$ are the eigenvectors of ρ, and $\hat{r}_i$ are estimates of the matching eigenvalues.

10.4 Toward Quantum Manifold Embedding

While quantum principal component analysis is attractive, more complex low-dimensional manifolds are also important for embedding high-dimensional data (Section 5.2). Isomap is an example that relies on local density: it approximates the geodesic distance between two points by the length of the shortest path between these two points on a neighborhood graph, then applies multidimensional scaling on the Gram matrix.

Constructing the neighborhood graph relies on finding the c smallest values of the distance function with high probability, using Grover's search. The overall complexity for finding the smallest values is $O(\sqrt{cN})$. This is equivalent to finding the smallest distances between a fixed point and the rest of the points in a set—that is, the neighbors of a given point (Aïmeur et al., 2013). Since there are a total of N points, the overall complexity of constructing the neighborhood graph is $O(N\sqrt{cN})$.

While calculating the geodesic distances is computationally demanding, the eigendecomposition of an $N \times N$ matrix in the multidimensional scaling is the bottleneck. The Gram matrix is always dense, and the same eigendecomposition applies as for quantum principal component analysis. Thus, for this step alone, an exponential speedup is feasible, provided that the input and output are quantum states.

10.5 Quantum *K*-Means

Most quantum clustering algorithms are based on Grover's search (Aïmeur et al., 2013). These clustering algorithms offer a speedup compared with their classical counterparts, but they do not improve the quality of the resulting clustering process. This is based on the belief that if finding the optimal solution for a clustering problem is NP-hard, then quantum computers would also be unable to solve the problem exactly in polynomial time (Bennett et al., 1997). If we use QRAM, an exponential speedup is possible.

The simplest quantum version of K-means clustering calculates centroids and assigns vectors to the closest centroids, like the classical variant, but using Grover's search to find the closest ones (Lloyd et al., 2013a). Since every vector is tested in each step, the complexity is $O(N \log(Nd))$.

Further improvement is possible if we allow the output to be quantum. Every algorithm that returns the cluster assignment for each N output must have at least $O(N)$ complexity. By allowing the result to be a quantum superposition, we remove this constraint. The output is a quantum state:

$$|\chi\rangle = (1/\sqrt{N}) \sum_j |c_j|j\rangle = (1/\sqrt{N}) \sum_c \sum_{j \in c} |c\rangle|j\rangle. \tag{10.16}$$

If necessary, state tomography on this reveals the clustering structure in classical terms. Constructing this state to ϵ accuracy has $O(\epsilon^{-1} K \log(KNd))$ complexity if we use the adiabatic theorem. If the clusters are well separated, the complexity is even less, $O(\epsilon^{-1} \log(KNd))$, as the adiabatic gap is constant (Section 4.3).

To construct the state in Equation 10.16, we start by selecting K vectors with labels i_c as initial seeds. We perform the first step of assigning the rest of the vectors to these clusters through the adiabatic theorem. We use the state

$$\frac{1}{\sqrt{NK}}\sum_{c'j}|c'\rangle|j\rangle\left(\frac{1}{\sqrt{K}}\sum_{c}|c\rangle|i_c\rangle\right)^{\otimes D}, \tag{10.17}$$

where the D copies of the state $\frac{1}{\sqrt{K}}\sum_c|c\rangle|i_c\rangle$ enable us to evaluate the distances $\|\mathbf{x}_j-\mathbf{x}_{i'_c}\|^2$ with the algorithm outlined in Section 10.2. The result is the initial clustering

$$|\psi_1\rangle=\frac{1}{\sqrt{N}}\sum_{c}\sum_{j\in c}|c\rangle|j\rangle, \tag{10.18}$$

where the states $|j\rangle$ are associated with the cluster c with the closest seed vector $\mathbf{x}_{i_c}$.

Assume that we have D copies of this state. Then, we can construct the individual clusters as $|\phi_1^c\rangle=(1/\sqrt{M_c})\sum_{k\in c}\mathbf{x}_k$, and thus estimate the number of states M_c in the cluster c.

With D copies of ψ_1, together with the clusters $|\phi_1^c\rangle$, we are able to evaluate the average distance between $\mathbf{x}_j$ and the mean of the individual clusters:

$$\left\|\mathbf{x}_j-\frac{1}{M_c}\sum_{k\in c}\mathbf{x}_k\right\|^2=\|\mathbf{x}_j-\mathbf{x}_c\|^2. \tag{10.19}$$

Next, we apply a phase to each component $|c'\rangle|j\rangle$ of the superposition with the following Hamiltonian:

$$H_{\mathrm{f}}=\sum_{c'j}\|\mathbf{x}_j-\mathbf{x}_{c'}\|^2|c'\rangle\langle c'|\otimes|j\rangle\langle j|\otimes I^{\otimes D}. \tag{10.20}$$

We start the adiabatic evolution on the state $(1/\sqrt{NK})\sum_{c'j}|c'\rangle|j\rangle|\psi_1\rangle^{\otimes D}$. The base Hamiltonian is $H_{\mathrm{b}}=I-|\phi\rangle\langle\phi|$, where $|\phi\rangle$ is the superposition of cluster centroids. The final state is

$$\left(\frac{1}{N}\sum_{c}\sum_{j\in c'}^{\prime}|c'\rangle|j\rangle\right)|\psi_1\rangle^{\otimes D}=|\psi_2\rangle|\psi_1\rangle^{\otimes D}. \tag{10.21}$$

Repeating this D times, we create D copies of the updated state $|\psi_2\rangle$. We repeat this cluster assignment step in subsequent iterations, eventually converging to the superposition $|\chi\rangle$ in Equation 10.16.

10.6 Quantum K-Medians

In K-means clustering, the centroid may lie outside the manifold in which the points are located. A flavor of this family of algorithms, K-medians, bypasses this problem by always choosing an element in a cluster to be the center. We achieve this by constructing a routine for finding the median in a cluster using Grover's search, and then iteratively calling this routine to reassign elements (Aïmeur et al., 2013).

Assume that we establish the distance between vectors either with the help of an oracle or with a method described in Section 10.2. To calculate the distance from a point $\mathbf{x}_i$ to all other points ($\sum_{j=1}^{N} d(\mathbf{x}_i, \mathbf{x}_j)$), we repeatedly call the distance calculator. Applying the modified Grover's search of Durr and Hoyer (1996) to find the minimum, we identify the median among a set of points in $O(N\sqrt{N})$ time.

We call this calculation of the median in each iteration of the quantum K- medians, and reassign the data points accordingly. The procedure is outlined in Algorithm 4.

ALGORITHM 4 Quantum K-medians

```
Require: Initial K points
Ensure: Clusters and their medians
  repeat
    for all x_i do
      Attach to the closest center
    end for
    for all K cluster do
      Calculate median for cluster
    end for
  until Clusters stabilize
```

10.7 Quantum Hierarchical Clustering

Quantum hierarchical clustering hinges on ideas similar to those of quantum K-medians clustering. Instead of finding the median, we use a quantum algorithm to calculate the maximum distance between two points in a set. We iteratively call this algorithm to split clusters and reassign the data instances to the most distant pair of instances (Aïmeur et al., 2013). This is the divisive form of hierarchical clustering.

As in Section 10.6, finding the maximum distance between a set of points relies on modifying the algorithm of Durr and Hoyer (1996). We initialize a maximum distance, $d_{\max}$, as zero. Then, we repeatedly call Grover's search to find indices i and j, such that $d(\mathbf{x}_i, \mathbf{x}_j) > d_{\max}$, and update the value of $d_{\max}$. The iterations stop when there are no new i, j index pairs.

Using the last pair, we attach every instance in the current cluster to a new cluster C_i or C_j, depending on whether the instance is closer to $\mathbf{x}_i$ or $\mathbf{x}_j$. Algorithm 5 outlines the process.

The same problems remain with the quantum variant of divisive clustering as with the classical one. It works if the clusters separate well and they are balanced. Outliers or a blend of small and large clusters will skew the result.

ALGORITHM 5 Quantum divisive clustering

```
Require: A cluster C
Ensure: Further clusters in C
  if Maximum depth reached then
     return  C
  end if
  Find most distant points x_i and x_j in C
  for all x ∈ C do
     Assign x to a new cluster C_i or C_j based on proximityx_i or x_j.
  end for
  Recursive call on C_i and C_j
```

10.8 Computational Complexity

The great speedups in quantum unsupervised learning pivot on quantum input and output states. Quantum principal component analysis with QRAM, if take some unit time for the evolution of the system, has a complexity of $O(\epsilon^{-1} \log d)$, where ϵ is the desired complexity; this is an exponential speedup over any known classical algorithm.

The quantum K-means algorithm with quantum input and output states has a complexity of $O(\log Nd)$, which is an exponential speedup over the polynomial complexity classical algorithms. Quantum K-medians has $O\left(\frac{tN^{3/2}}{\sqrt{K}}\right)$ complexity as opposed to $O\left(\frac{tN^2}{K}\right)$ of the classical algorithm, where t is the number of iterations. Quantum divisive clustering, $O(N \log N)$, has much lower complexity compared with the classical limit $\Theta(N^2)$. We must point out, however, that classical sublinear clustering in a probably approximately correct setting exists, using statistical sampling. This method reduces the complexity to $O(\log^2 N)$ (Mishra et al., 2001). Sublinear clustering algorithms are not exclusive to quantum systems.

Quantum Pattern Recognition

11

Artificial neural networks come in countless variants, and it is not surprising that numerous attempts aim at introducing quantum behavior into these models. While neurons and connections find analogues in quantum systems, we can build models that mimic classical neural networks, but without using the neural metaphor. For instance, a simple scheme to train the quantum variant of Hopfield networks is to store patterns in a quantum superposition; this is the model of quantum associative memories. To match a pattern, we use a modified version of Grover's search. Storage capacity is exponentially larger than in classical Hopfield networks (Section 11.1).

Other models do include references to classical neural networks. Neurocomputing, however, is dependent on nonlinear components, but evolution in quantum systems is linear. Measurement is a key ingredient in introducing nonlinearity, as the quantum perceptron shows (Section 11.2).

Apart from measurement, decoherence paves the way to nonlinear patterns, and Feynman's path integral formulation is also able to introduce nonlinearity. Feedforward networks containing layers and multiple neurons may mix quantum and classical components, introducing various forms of nonlinear behavior (Section 11.3).

Multiple attempts show the physical feasibility of implementing quantum neural networks. Nuclear magnetic resonance and quantum dot experiments are promising, albeit using only a few qubits. There are suggestions for optical and adiabatic implementations (Section 11.4).

While we are far from implementing deep learning networks with quantum components, the computational complexity is, not surprisingly, much better than with classical algorithms (Section 11.5).

11.1 Quantum Associative Memory

A quantum associative memory is analogous to a Hopfield network (Section 6.2), although the quantum formulation does not require a reference to artificial neurons. A random access memory retrieves information using addresses (Section 10.1), whereas recall from an associative memory requires an input pattern instead of an address. The pattern may be a complete match to one of the patterns stored in the associative memory, but it can also be a partial pattern. The task is to reproduce the closest match.

Quantum Machine Learning. http://dx.doi.org/10.1016/B978-0-12-800953-6.00001-1

If the dimension of a data instance is d, and we are to store N patterns, a classical Hopfield network would require d neurons and the number of patterns that can be stored is linear in d (Section 6.2). The quantum associative memory stores patterns in a superposition, offering a storage capacity of $O(2^d)$, using only d neurons (Ventura and Martinez, 2000). Here we follow the exposition in Trugenberger (2001), which extends and simplifies the original formulation by Ventura and Martinez (2000).

The associative memory is thus a superposition with equal probability for all of its entangled qubits:

$$|M\rangle = \frac{1}{\sqrt{N}} \sum_{i=1}^{N} |x_i\rangle, \tag{11.1}$$

where each $|x_i\rangle$ state has d qubits. The first task is to construct the superposition $|M\rangle$. We use auxiliary registers to do so.

The first register x of d qubits will temporarily hold the subsequent data instances x_i. The second register is a two-qubit utility register u. The third register is the memory m of d qubits. With these registers, the initial quantum state is

$$|\psi_0^1\rangle = |x_{11}, \ldots, x_{1d}; 01; 0, \ldots, 0\rangle. \tag{11.2}$$

The upper index of $|\psi_0^1\rangle$ is the current data instance being processed, and the lower index is the current step in storing the data instance. There will be a total of seven steps.

We separate this state into two terms, one corresponding to the patterns already in the memory, and the other processing a new pattern. The state of the utility qubit distinguishes the two parts: $|0\rangle$ for the stored patterns, and $|1\rangle$ for the part processed.

To store $\mathbf{x}_i$, we copy the pattern into the memory register with

$$|\psi_1^i\rangle = \prod_{j=1}^{d} 2\text{XOR}_{x_{ij}u_2m_j} |\psi_0^i\rangle. \tag{11.3}$$

Here 2XOR is a Toffoli gate (Section 4.2), and the lower index indicates which qubits it operates on.

Then we flip all qubits of the memory register to $|1\rangle$ if the contents of the pattern and the memory registers are identical, which is true only for the processing term:

$$|\psi_2^i\rangle = \prod_{j=1}^{d} \text{NOT}_{m_j}\text{XOR}_{x_{ij}m_j} |\psi_1^i\rangle. \tag{11.4}$$

The third operation changes the first utility qubit to $|1\rangle$:

$$|\psi_3^i\rangle = n\text{XOR}_{m_1 \cdots m_d u_1} |\psi_2^i\rangle. \tag{11.5}$$

The fourth step uses a two-qubit gate:

$$CS^i = \text{diag}(I, S^i), \tag{11.6}$$

where

$$S^i = \begin{pmatrix} \sqrt{\frac{i-1}{i}} & \frac{1}{\sqrt{i}} \\ -\frac{1}{\sqrt{i}} & \sqrt{\frac{i-1}{i}} \end{pmatrix}. \tag{11.7}$$

With this gate, we separate the new pattern to be stored, together with the correct normalization factor:

$$|\psi_4^i\rangle = CS_{u_1u_2}^{N+1-i}|\psi_3^i\rangle. \tag{11.8}$$

We perform the reverse operations of Equations 11.4 and 11.5, first restoring the utility qubit u_1,

$$|\psi_5^i\rangle = n\text{XOR}_{m_1\cdots m_d u_1}|\psi_4^i\rangle, \tag{11.9}$$

and then restoring the memory register m to their original values,

$$|\psi_6^i\rangle = \prod_{j=d}^{1} \text{XOR}_{x_{ij}m_j}\text{NOT}_{m_j}|\psi_5^i\rangle. \tag{11.10}$$

After this step, we have the following state:

$$|\psi_6^i\rangle = \frac{1}{\sqrt{N}}\sum_{k=1}^{i}|x_i;00;x_k\rangle + \sqrt{\frac{N-i}{N}}|p^i;01;p^i\rangle. \tag{11.11}$$

In the last step, we restore the third register in the second term of Equation 11.11 to $|0\rangle$:

$$|\psi_7^i\rangle = \prod_{j=d}^{1} 2\text{XOR}_{x_{ij}u_2m_j}|\psi_6^i\rangle. \tag{11.12}$$

A second look at the superposition state $|M\rangle$ in Equation 11.1 highlights an important characteristic: the stored states have equal weights in the superposition. Grover's algorithm works on such uniformly weighted superpositions (Equation 4.25). Yet, if we wish to retrieve a state given an input, the original Grover's algorithm as outlined in Section 4.5 will not work efficiently: the probability of retrieving the correct item will be low. Furthermore, Grover's algorithm assumes that all patterns of a given length of qubits are stored, which is hardly the case in pattern recognition.

We modify Algorithm 1 to overcome this problem (Ventura and Martinez, 2000). The modified variant is outlined in Algorithm 6. Zhou and Ding (2008) suggested a similar improvement.

The second state rotation operator rotates the phases of the desired states, moreover, and it rotates the phases of all the stored patterns as well. The states not matching the target pattern are present as noise, and the superposition created by applying Hadamard transforms also introduces undesirable states. The extra rotations in the modified algorithm force these two different kinds of nondesired states to have the same phase, rather than opposite phases as in the original algorithm. The eventual state then serves as the input to the normal loop of Grover's algorithm.

ALGORITHM 6 Grover's algorithm modified for pattern recognition

```
Require: Initial state in equal superposition |M⟩ = 1/√N ∑_{i=1}^N |x_i⟩, the
  input state to be matched through a corresponding oracle O.
Ensure: Most similar element retrieved.
  Apply the Grover operator on the memory state: |M⟩ = G|M⟩.
  Apply an oracle operator for the entire stored memory state
  |M⟩ = O_M|M⟩.
  Apply the rest of the Grover operator: |M⟩ = H^{⊗n}(2|0⟩⟨0| − I)H^{⊗n}|M⟩.
  for O(√N) times do
    Apply the Grover operator G
  end for
  Measure the system.
```

A similar strategy is used in quantum reinforcement learning. Distinct from supervised learning, input and output pairs are associated with a reward, and the eventual goal is to take actions that maximize the reward. Actions and states are stored in a superposition, and Grover's search amplifies solutions that have a high reward (Dong et al., 2008).

To find a matching pattern in the quantum associative memory, we perform a probabilistic cloning of the memory state $|M\rangle$, otherwise we would lose the memory state after retrieving a single item (Duan and Guo, 1998). The number of elements stored can be exponential, which makes the time complexity of Grover's algorithm high.

An alternative method splits retrieval into two steps: identification and recognition (Trugenberger, 2001). We must still have a probabilistic clone of the memory state. We use three registers in the retrieval process: the first register contains the input pattern, the second register stores $|M\rangle$, and there is a control register c initialized to the state $(|0\rangle + |1\rangle)/\sqrt{2}$. The initial state for the retrieval process is thus

$$\begin{aligned}|\psi_0\rangle = &\frac{1}{\sqrt{2N}}\sum_{k=1}^{N}|i_1,\ldots,i_n;x_{k1},\ldots,x_{kn};0\rangle\\ &+\frac{1}{\sqrt{2N}}\sum_{k=1}^{N}|i_1,\ldots,i_n;x_{k1},\ldots,x_{kn};1\rangle.\end{aligned} \tag{11.13}$$

We flip the memory registers to $|1\rangle$ if i_j and x_{kj} are identical:

$$|\psi_1\rangle = \prod_{k=1}^{d}\mathrm{NOT}_{m_k}\mathrm{XOR}_{i_k m_k}|\psi_0\rangle. \tag{11.14}$$

We want to calculate the Hamming distance (Equation 2.3) between the input pattern and the instances in the memory. To achieve this, we use the following Hamiltonian:

$$H = d_m \otimes (\sigma_z)_c,$$

$$d_m = \sum_{k=1}^{n} \left(\frac{\sigma_z + 1}{2} \right)_{m_k},$$

where σ_z is the third Pauli matrix (Equation 4.3). This Hamiltonian measures the number of 0's in register m with a plus sign if c is $|0\rangle$, and the number with a minus sign if c is in $|1\rangle$.

The terms of the superposition in$|\psi_1\rangle$ are eigenstates of H. Applying the corresponding unitary evolution (Section 3.41), we get

$$\begin{aligned} |\psi_2\rangle = &\frac{1}{\sqrt{2N}} \sum_{k=1}^{N} \mathrm{e}^{\mathrm{i}(\pi/2d)d(i,x_k)} |i_1, \ldots, i_d; b_{k1}, \ldots, b_{kd}; 0\rangle \\ &+ \frac{1}{\sqrt{2N}} \sum_{k=1}^{N} \mathrm{e}^{-\mathrm{i}(\pi/2d)d(i,x_k)} |i_1, \ldots, i_d; b_{k1}, \ldots, b_{kd}; 1\rangle, \end{aligned} \tag{11.15}$$

where $b_{kj} = 1$ if and only if $i_j = x_{kj}$.

As the last step, we perform the inverse operation of Equation 11.14:

$$|\psi_3 = H_\mathrm{c} \prod_{k=d}^{1} \mathrm{XOR}_{i_k m_k} \mathrm{NOT}_{m_k} |\psi_2\rangle, \tag{11.16}$$

where H_c is the Hadamard gate on the control qubit. We measure the system in the control qubit, obtaining the probability distribution:

$$\mathbf{P}(|c\rangle = |0\rangle) = \sum_{k=1}^{N} \frac{1}{N} \cos^2 \left(\frac{\pi}{2d} d(i, x_k) \right), \tag{11.17}$$

$$\mathbf{P}(|c\rangle = |1\rangle) = \sum_{k=1}^{N} \frac{1}{N} \sin^2 \left(\frac{\pi}{2d} d(i, x_k) \right). \tag{11.18}$$

Thus, if the input pattern is substantially different from every stored pattern, the probability is higher of measuring $|c\rangle = |1\rangle$. If the input pattern is close to every stored pattern, the probability of measuring $|c\rangle = 0$ will be higher. Repeating the algorithm gives an improved estimate.

Once a pattern has been recognized, we proceed to measure the memory register to identify the closest state. The probability of obtaining a pattern x_k is

$$\mathbf{P}(x_k) = \frac{1}{N\mathbf{P}(|c\rangle = |0\rangle)} \cos^2 \left(\frac{\pi}{2d} d(i, x_k) \right). \tag{11.19}$$

The probability peaks around the patterns which have the smallest Hamming distance to the input.

Table 11.1 provides a high-level comparison of classical Hopfield networks and quantum associative memories.

The recognition efficiency relies on comparing cosines and sines of the same distances in the distribution, whereas the identification efficiency relies on comparing cosines of the different distances in the distribution. Identification is most efficient

Table 11.1 Comparison of Classical Hopfield Networks and Quantum Associative Memories (Based on Ezhov and Ventura, 2000)

Classical Hopfield Network	Quantum Associative Memory
Connections w_{ij}	Entanglement $\lvert x_{k1} \cdots x_{kd}\rangle$
Learning rule $\sum_{k=1}^{N} x_{ki}x_{kj}$	Superposition of entangled states $\sum_{k=1}^{N} a_k\lvert x_{k1} \cdots x_{kd}\rangle$
Winner search $n = \operatorname{argmax}_i(f_i)$	Unitary transformation $\lvert\psi\rangle \to \lvert\psi'\rangle$
Output result n	Decoherence $\sum_{k=1}^{N} a_k\lvert\mathbf{x}_k\rangle \Rightarrow \lvert\mathbf{x}_n\rangle$

when one of the distances is zero, and all others are large, making the probability peak on a single pattern. This is the opposite scenario for optimal recognition efficiency, which prefers uniformly distributed distances. Increasing the number of control qubits from 2 to d improves the efficiency of identification (Trugenberger, 2002).

11.2 The Quantum Perceptron

As in the case of classical neural networks, the perceptron is the simplest model of a feedforward model. The quantum variant of a perceptron relies on the unitary evolution of a density matrix (Lewenstein, 1994). The quantum perceptron takes an input state ρ^{in}, and produces an output $\rho^{\text{out}} = U\rho^{\text{in}}U^\dagger$ (see Equation 3.42). The output density matrix is then subject to measurement to introduce nonlinearity.

Given N patterns to learn, we prepare the states ρ_i^{in}, $i = 1,\ldots,N$ with projection operators P_i^{in}. If ρ_0 is the density matrix, the preparation of the system consists of stating the ith input question—that is, the measurement and renormalization

$$\rho_i^{\text{in}} = P_i^{\text{in}}\rho_0 P_i^{\text{in}}/\text{tr}(P_i^{\text{in}}\rho_0 P_i^{\text{in}}). \tag{11.20}$$

The output states are defined similarly with corresponding projection operators P_i^{out}. These measurements ensure a nonlinear behavior in the system.

Let us define two local cost functions on the individual data instances. Let E_i denote the conditional probability that we did not find the state in the ith state, although it was in the ith state:

$$E_i = \text{tr}(Q_i^{\text{out}}UP_i^{\text{in}}\rho_0 P_i^{\text{in}}U^\dagger Q_i^{\text{out}})/\text{tr}(P_i^{\text{in}}\rho_0 P_i^{\text{in}}), \tag{11.21}$$

where $Q = 1 - P$. Analogously, F_i is the conditional probability of finding the system in the ith state, whereas it was not in the ith state:

$$F_i = \text{tr}(P_i^{\text{out}}UQ_i^{\text{in}}\rho_0 Q_i^{\text{in}}U^\dagger P_i^{\text{out}})/\text{tr}(Q_i^{\text{in}}\rho_0 Q_i^{\text{in}}). \tag{11.22}$$

With these two measures of error, we define a global cost function

$$E = \sum_{i=1}^{N} \frac{E_i + F_i}{2N}. \tag{11.23}$$

Rather than asking which unitary transformation U would minimize the cost function in Equation 11.23, which would be similar to the task set out in Chapter 13 relating quantum process tomography, we are interested in restricting the set of unitaries, and asking what is the probability of finding one given an accepted probability of error (Gardner, 1988).

Assume that the inputs and outputs are independent, and that the P_i^{in} operators are one-dimensional. Furthermore, assume that the P_i^{out} operators are D-dimensional. The ratio D/d will determine the learning capacity of the quantum perceptron.

When the ratio D/d is too large, the constraint on the error bound on F_i cannot be satisfied, and we have a trivial model that gives the answer yes to any input—that is, the perceptron does not perform any computation.

If E_i and F_i are bounded by a and a', respectively, and D/d is between a' and $1-a$, we always obtain a satisfactory response for any choice of U. In this case, one of the error terms must exceed $1/2$; hence, practically no information is stored in the perceptron.

As the ratio D/d becomes smaller, we approximate a more standard form of learning. The error cannot be arbitrarily small, as it is bounded from below by

$$D/d \geq \frac{3b^2}{\left(1+b+\sqrt{1+b}\right)^2 - b^2}, \tag{11.24}$$

where $b = 3(1-a)$. Finding an optimal U is possible, but Lewenstein (1994) does not show a way of doing this. As we are not concerned with additional restrictions on the unitary, theoretically the quantum perceptron has no limits on its storage capacity.

11.3 Quantum Neural Networks

Classical networks have nonlinear irreversible dynamics, whereas quantum systems evolve in a linear, reversible way (Zak and Williams, 1998). How do we scale up nonlinearity from a single perceptron to a network of neurons?

If we keep implementations in mind, quantum dots are a candidate for quantum neural networks (Section 11.4). This allows nonlinearity to enter through Feynman's path integral formulation (Behrman et al., 1996).

A commoner approach is to alternate classical and quantum components, in which the measurement and the quantum collapse introduce nonlinearity, as in the case of a single perceptron (Narayanan and Menneer, 2000; Zak and Williams, 1998). In fact, numerical simulations showed that a fully quantum neural network may produce worse results than a partly quantum one (Narayanan and Menneer, 2000). After each measurement, the quantum device is reset to continue from its eigenstate. To overcome the probabilistic nature of the measurements, several quantum devices are measured and reset simultaneously (Zak and Williams, 1998).

The overall architecture of the network resembles the classical version, mixing layers of classical and quantum components (Narayanan and Menneer, 2000). Input nodes are replaced by slits through which quantum particles can travel. The particles

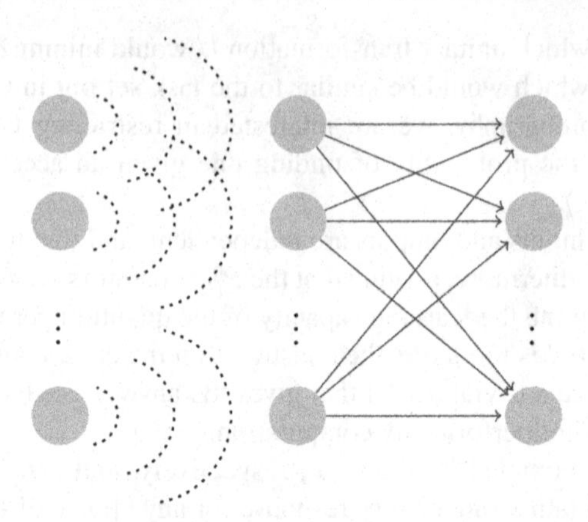

Figure 11.1 In a feedforward quantum neural network, quantum and classical components may mix. In this figure, the connections in the first and second layers are replaced by interference, but the edges between the second and third layers remain classical.

undergo some unitary evolution, and an interference pattern is drawn on a measurement device (Figure 11.1). This interference pattern replaces weights, and this pattern is modified by Grover's algorithm through the training procedure.

The learning capacity of quantum neural networks is approximately the same as that of the classical variants (Gupta and Zia, 2001). Their advantage comes from a greater generalization performance, especially when training has low error tolerance, although these results are based on numerical experiments on certain data sets (Narayanan and Menneer, 2000).

11.4 Physical Realizations

Quantum dot molecules are nearby groups of atoms deposited on a host substrate. If the dots sufficiently close to one another, excess electrons can tunnel between the dots, which gives rise to a dipole. This is a model for a qubit, and, since it is based on solid-state materials, it is an attractive candidate for implementations. Quantum dots are easy to manipulate by optical means, changing the number of excitations. This leads to a temporal neural network: temporal in the sense nodes are successive time slices of the evolution of a single quantum dot (Behrman et al., 2000). If we allow a spatial configuration of multiple quantum dots, Hopfield networks can be trained.

Optical realizations have also been suggested. Lewenstein (1994) discussed two potential examples for implementing perceptrons, a d-port lossless linear optical unit, and a d-port nonlinear unit. In a similar vein, Altaisky (2001) mooted phase shifters and beam splitters for linear evolution, and light attenuators for the nonlinear case. A double-slit experiment is a straightforward way to implement the interference model of feedforward networks (Narayanan and Menneer, 2000). Hopfield networks have a holographic model implementation (Loo et al., 2004).

Adiabatic quantum computing offers a global optimum for quantum associative memories, as opposed to the local optimization in a classical Hopfield network (Neigovzen et al., 2009). A two-qubit implementation was demonstrated on a liquid-state nuclear magnetic resonance system. A quantum neural network of N bipolar states is represented by N qubits. Naturally, a superposition of "fire" and "not fire" exists in the qubits. The Hamiltonian is given by

$$H_{\mathrm{p}} = H_{\mathrm{mem}} + \Gamma H_{\mathrm{inp}}, \tag{11.25}$$

where H_{mem} represents the knowledge of the stored pattern in the associative memory, H_{inp} represents the computational input, and $\Gamma > 0$ is an appropriate weight.

The memory Hamiltonian is defined as the coupling strengths between qubits:

$$H_{\mathrm{mem}} = -\frac{1}{2}\sum_{i \neq j} w_{ij}\sigma_i^z \sigma_j^z, \tag{11.26}$$

where σ_i^z is the Pauli Z matrix on qubit i, and w_{ij} are the weights of the Hopfield network.

For the retrieval Hamiltonian H_{inp}, it is assumed that the input pattern is of length N. If it is not, we pad the missing states with zero. The term is defined as

$$H_{\mathrm{inp}} = \sum_j |x_{ij}\rangle \sigma_j^z. \tag{11.27}$$

The external field defined by H_{inp} creates a metric that is proportional to the Hamming distance between the input state and the memory patterns. If the energy of the memory Hamiltonian H_{mem} is shifted, similar patterns will have lower energy (Figure 11.2).

This scheme ignores training: it assumes that the memory superposition contains all configurations. This is in contrast with the learning algorithm described in Section 11.1.

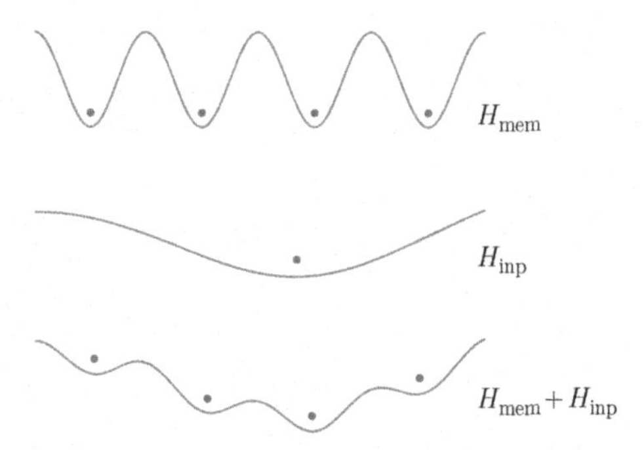

Figure 11.2 In a quantum associative memory that relies on the adiabatic theorem, the patterns are stored in the stable states of the Hamiltonian H_{mem}. The input pattern is represented by a new Hamiltonian H_{inp}, changing the overall energy landscape, $H_{\mathrm{mem}} + H_{\mathrm{inp}}$. The ground state of the composite system points to the element to be retrieved from the memory.

11.5 Computational Complexity

Quantum associative memory has a linear time complexity in the number of elements to be folded in the superposition and the number of dimensions ($O(Nd)$; Ventura and Martinez, 2000). Assuming that $N \gg d$, this means that computational complexity is improved by a factor of N compared with the classical Hopfield network.

Quantum artificial neural networks are most useful where the training set is large (Narayanan and Menneer, 2000). Since the rate of convergence of the gradient descent in a classical neural network is not established for the general case, it is hard to estimate the overall improvement in the quantum case. The generic improvement with Grover's algorithm is quadratic, and we can assume this for the quantum components of a quantum neural network.

Quantum Classification

12

Many techniques that made an appearance in Chapter 10 are useful in the supervised setting. Quantum random access memory (QRAM), the calculation of the kernel matrix, and the self-analysis of states provide the foundations for this chapter.

The quantum version of nearest-neighbor classification relies either on a QRAM or on oracles. Apart from extending the case of the quantum K-means to labeled instances, an alternative variant compares the minimum distances across all elements in the clusters, yielding better performance (Section 12.1).

A simple formulation of quantum support vector machines uses Grover's search to replace sequential minimum optimization in a discretized search space (Section 12.2).

Least-squares support vector machines translate an optimization problem into a set of linear equations. The linear equations require the quick calculation of the kernel matrix—this is one source of the speedup in the quantum version. The other source of the speedup is the efficient solution of the linear equations on quantum hardware. Relying on this formulation, and assuming quantum input and output space, quantum support vector machines can achieve an exponential speedup over their classical counterparts (Section 12.3).

Generic computational complexity considerations are summarized at the end of the chapter (Section 12.4).

12.1 Nearest Neighbors

Nearest neighbors is a supervised cluster assignment task. Assume that there are two clusters, and they are represented by the centroid vector $(1/M_c)\sum_{j\in c}\mathbf{x}_j$, where c is the class. The task is to find the minimum distance for a new vector $\mathbf{x}$ such that the distance $|\mathbf{x}-(1/M_c)\sum_{j\in c}\mathbf{x}_j|^2$ is minimal. We use the algorithm in Section 10.2 to estimate this distance with a centroid vector. We construct this centroid state by querying the QRAM (Section 10.1).

An extension of this algorithm avoids using the centroid vector, as this form of nearest-neighbor classification performs poorly if the classes do not separate well, or if the shape of the classes is complicated and the centroid does not lie within the class (Wiebe et al., 2014). Instead, we can evaluate the distances between all vectors in a class and the target vector, and the task becomes deciding whether the following inequality is true:

Quantum Machine Learning. http://dx.doi.org/10.1016/B978-0-12-800953-6.00001-1

$$\min_{\mathbf{x}_j \in c_1} \|\mathbf{x} - \mathbf{x}_j\| \leq \min_{\mathbf{x}_j \in c_2} \|\mathbf{x} - \mathbf{x}_j\|. \tag{12.1}$$

For this variant, we assume that the vectors are f-sparse—the vectors do not contain more than f nonzero entries. In many classical algorithms, execution speed—and not complexity—is improved by exploiting sparse data structures. Such structures are common in applications such as text mining and recommendation engines. We also assume that there is an upper bound $r_{\max}$ on the maximum value of any feature in the data set. We further assume that we have two quantum oracles:

$$\begin{aligned} \mathcal{O}|j\rangle|i\rangle|0\rangle &:= |j\rangle|i\rangle|x_{ji}\rangle, \\ \mathcal{F}|j\rangle|l\rangle &:= |j\rangle|f(j,l)\rangle, \end{aligned} \tag{12.2}$$

where $f(j, l)$ gives the location of the lth nonzero entry in $\mathbf{x}_j$.

Given these conditions, finding the maximum overlap, $\max_j |\langle \mathbf{x}|\mathbf{x}_j\rangle|^2$, within error at most ϵ and with success probability at least $1 - \delta$ requires an expected number of queries that is on the order of

$$O\left(\frac{\sqrt{N} f^2 r_{\max}^4 \ln\left(\frac{N \ln(N)}{\delta}\right)}{\epsilon}\right). \tag{12.3}$$

To achieve this bound, we use a swap test (Section 10.2) on the following states:

$$\begin{aligned} &\frac{1}{\sqrt{f}} \sum_i |i\rangle \left(\sqrt{1 - \frac{r_{ji}^2}{r_{\max}^2}} e^{-i\phi_{ji}}|0\rangle + \frac{x_{ji}}{r_{\max}}|1\rangle\right)|1\rangle, \\ &\frac{1}{\sqrt{f}} \sum_i |i\rangle|1\rangle \left(\sqrt{1 - \frac{r_{0i}^2}{r_{\max}^2}} e^{-i\phi_{0i}}|0\rangle + \frac{x_{0i}}{r_{\max}}|1\rangle\right), \end{aligned} \tag{12.4}$$

where r_{ji} comes from the polar form of the number $x_{ji} = r_{ji}e^{i\phi_{ji}}$. These states are prepared by six oracle calls and two single-qubit rotations (Wiebe et al., 2014).

We would like to determine the probability of obtaining 0. To get a quick estimate, we do not perform a measurement in the swap test, but apply amplitude estimation instead to achieve a scaling of $O(1/\epsilon)$ in estimating $\mathbf{P}(0)$ with ϵ accuracy. Let us denote this estimate by $\hat{\mathbf{P}}(0)$. Given a register of dimension R, the error of this estimation is bounded by

$$|\mathbf{P}(0) - \hat{\mathbf{P}}(0)| \leq \frac{\pi}{R} + \frac{\pi^2}{R^2}. \tag{12.5}$$

This way, R must be at least $R \geq \lceil 4\pi(\pi + 1) f^2 r_{\max}^4/\epsilon \rceil$ to achieve an error bound of $\epsilon/2$. Once we have an estimate of $\mathbf{P}(0)$, then the overlap is given by

$$|\langle \mathbf{x}|\mathbf{x}_j\rangle|^2 = (2\mathbf{P}(0) - 1) f^2 r_{\max}^4. \tag{12.6}$$

A similar result holds for the Euclidean distance.

An alternative view on nearest-neighbor clustering is to treat the centroid vectors as template states, and deal with the problem as quantum template matching. Quantum template matching resembles various forms of state tomography. In quantum state discrimination, we know the prior distribution of a set of states $\{\rho_i\}$, and the task is to decide which state we are receiving. In quantum state estimation, we reconstruct a given unknown state by estimating its parameters. In template matching, we have a prior set of states, as in state discrimination, but the unknown state is also characterized by parameters, as in state estimation (Sasaki and Carlini, 2002; Sasaki et al., 2001).

To estimate the distance between the unknown state and the target template state, we use fidelity (Section 4.7), and we choose the template for which the fidelity is the largest. The matching strategy is represented by a positive operator–valued measure, $\{P_j\}$, and its performance is measured by its average score over all matches. The scenario applies to both classical and quantum inputs. In the first setting, the template states are derived from classical information. In the second setting, they generalize to fully quantum template states, where only a finite number of copies of each template state are available. Yet, a generic strategy for finding the optimal positive operator–valued measure, does not exist, and we can deal only with special cases in which the template states have a specific structure.

12.2 Support Vector Machines with Grover's Search

The simplest form of quantum support vector machines observes that if the parameters of the cost function are discretized, we can perform an exhaustive search in the cost space (Anguita et al., 2003). The search for the minimum is based on a variant of Grover's search (Durr and Hoyer, 1996).

The idea's simplicity is attractive: there is no restriction on the objective function. Since we do not depend on an algorithm like gradient descent, the objective function might be nonconvex. The real strength of quantum support vector machines in this formulation might be this ability to deal with nonconvex objective functions, which leads to better generalization performance, especially if outliers are present in the training data.

As pointed out in Section 9.3, a convex loss function applies a large penalty to points with large negative margin—these are points which are incorrectly classified and are far from the boundary. This is why outliers and classification noise affect a convex loss function. Carefully designed nonconvex loss functions may avoid this pitfall (Section 9.4).

Since adiabatic quantum optimization is ideal for such nonconvex optimization (Denchev et al., 2012), it would be worth casting support vector machines as a quadratic unconstrained binary optimization program (Section 14.2), suitable for an adiabatic quantum computer.

This formulation leaves calculating the kernel matrix untouched, and it has $O(N^2d)$ complexity. Although the search process has a reduced complexity, calculations will be dominated by generating the kernel matrix.

12.3 Support Vector Machines with Exponential Speedup

If we calculate only the kernel matrix by quantum means, we have a complexity of $O(M^2(M+\epsilon^{-1}\log N))$. There is more to gain: exponential speedup in the number of training examples is possible when using the least-squares formulation of support vector machines (Section 7.5).

The algorithm hinges on three ideas:

- Quantum matrix inversion is fast (Harrow et al., 2009).
- Simulation of sparse matrixes is efficient (Berry et al., 2007),
- Nonsparse density matrices reveal the eigenstructure exponentially faster than in classical algorithms (Lloyd et al., 2013b).

To solve the linear formulation in Equation 7.28, we need to invert

$$F=\begin{pmatrix} 0 & \mathbf{1}^{\top} \\ \mathbf{1} & K+\gamma^{-1}I \end{pmatrix}. \tag{12.7}$$

The matrix inversion algorithm needs to simulate the matrix exponential of F. We split F as $F=J+K_{\gamma}$, where

$$J=\begin{pmatrix} 0 & \mathbf{1}^{\top} \\ \mathbf{1} & 0 \end{pmatrix} \tag{12.8}$$

is the adjacency matrix of a star graph, and

$$K_{\gamma}=\begin{pmatrix} 0 & 0 \\ 0 & K+\gamma^{-1}I \end{pmatrix}. \tag{12.9}$$

We normalize F with its trace: $\hat{F}=\frac{F}{\mathrm{tr}(F)}=\frac{F}{\mathrm{tr}(K_{\gamma})}$.

By using the Lie product formula, we get the exponential as

$$\mathrm{e}^{-\mathrm{i}\hat{F}\Delta t}=\mathrm{e}^{-\mathrm{i}\,J\Delta t/\mathrm{tr}(K_{\gamma})}\mathrm{e}^{-\mathrm{i}\gamma^{-1}I\Delta t/\mathrm{tr}(K_{\gamma})}\mathrm{e}^{-\mathrm{i}K\Delta t/\mathrm{tr}(K_{\gamma})}+O(\Delta t^2). \tag{12.10}$$

To obtain the exponentials, the sparse matrices J and the constant multiply of the identity matrix are easy to simulate (Berry et al., 2007).

On the other hand, the kernel matrix K is not sparse. This is where quantum self-analysis helps: given multiple copies of a density matrix ρ, it is possible to perform $\mathrm{e}^{-\mathrm{i}\rho t}$; this resembles quantum principal component analysis (Section 10.3). With the quantum calculation of the dot product and access to a QRAM (Sections 10.1 and 10.2), we obtain the normalized kernel matrix

$$\hat{K}=\frac{K}{\mathrm{tr}(K)}=\frac{1}{\mathrm{tr}(K)}\sum_{i,j=1}^{N}\langle \mathbf{x}_i|\mathbf{x}_j\rangle\,\|\mathbf{x}_i\|\|\mathbf{x}_j\|\,|i\rangle\langle j|. \tag{12.11}$$

$\hat{K}$ is a normalized Hermitian matrix, which makes it a prime candidate for quantum self analysis. The exponentiation is done in $O(\log N)$ steps.

We use Equation 12.10 to perform quantum phase estimation to get the eigenvalues and eigenvectors. If the desired accuracy is ϵ, we need $n=O(\epsilon^{-3})$ copies of the state.

Then, we express the **y** in Equation 7.28 in the eigenbasis, and invert the eigenvalues to obtain the solution $|b, \alpha\rangle$ of the linear equation (Harrow et al., 2009).

With this inversion algorithm, the overall time complexity of training the support vector parameters is $O(\log(Nd))$.

The kernel function is restricted. Linear and polynomial kernels are easy to implement; in fact, the inner product is evaluated directly in the embedding space. Radial basis function kernels are much harder to imagine in this framework, and these are the comment kernels. The approximation methods for calculating the kernel matrix apply (Section 10.2).

$O(\log(MN))$ states are required to perform classification. This is advantageous because it compresses the kernel exponentially. Yet, it is also a disadvantage because the trained model is not sparse. A pivotal point in the success of support vector machines is structural risk minimization: the learned model should not overfit the data, otherwise its generalization performance will be poor. The least-squares formulation and its quantum variant are not sparse: every data instance will become a support vector. None of the α_i values will be zero.

12.4 Computational Complexity

Equation 12.3 indicates that quantum nearest neighbors is nearly quadratically better than the classical analogue even with sparse data. This is remarkable as most quantum learning algorithms do not assume a sparse structure in the quantum states.

Since there are no strict limits on the rate of convergence of the gradient descent in the optimization phase of support vector machines, it is difficult to establish how much faster we can get if we use Grover's algorithm to perform this stage. The bottleneck in this simple formulation remains the $O(N^2d)$ complexity of calculating the kernel matrix.

If we rely on quantum input and output states, and replace the calculation of the kernel matrix with the method outlined in Section 10.2, we can achieve an exponential speedup over classical algorithms. Using the least-squares formulation of support vector machines, the self-analysis of quantum states will solve the matrix inversion problem, and the overall complexity becomes $O(\log(Nd))$.

Quantum Process Tomography and Regression

13

Symmetry is essential to why some quantum algorithms are successful—quantum Fourier transformation and Shor's algorithm achieve an exponential speedup by exploiting symmetries. A group can be represented by sets of unitary matrices with the usual multiplication rule—that is, sets of elements with a binary operation satisfying algebraic conditions. Groups are essential in describing symmetries, and since no such representation is possible using stochastic matrices on classical computers, we see why symmetries are so important in quantum computing. Quantum process tomography is a procedure to characterize the dynamics of a quantum system: symmetry and the representation theory of groups play an important role in it, which in turn leads to the efficient learning of unknown functions.

The task of regression thus translates well to quantum process tomography. The dynamic process which we wish to learn is a series of unitary transformations, also called a channel. If we denote the unitary by U, the goal becomes to derive an estimate U such that $U(\rho_{\text{in}}) = \rho_{\text{out}}$. Then, as in the classical case, we wish to calculate $U(\rho)$ for a new state ρ that we have not encountered before.

In a classical setting, we define an objective function, and we seek an optimum subject to constraints and assumptions. The assumption in learning by quantum process tomography is that the channel is unitary and that the unitary transformation is drawn from a group—that is, it meets basic symmetry conditions. The objective function is replaced by the fidelity of quantum states.

Apart from these similarities, the rest of the learning process does not resemble the classical variant. Unlike in the classical setting, learning a unitary requires a double maximization: we need an optimal measuring strategy that optimally approximates the unitary, and we need an optimal input state that best captures the information of the unitary (Acín et al., 2001). This resembles active learning, where we must identify data instances that improve the learned model (Section 2.3).

The key steps are as follows:

- Storage and parallel application of the unitary on a suitable input state that achieve optimum storage (Section 13.5).
- A superposition of maximally entangled states is the optimal input state (Section 13.6).
- A measure-and-prepare strategy on the ancilla with an optimal positive operator–valued measure (POVM) is best for applying the learned unitary on an arbitrary number of new states (Section 13.7).

Quantum Machine Learning. http://dx.doi.org/10.1016/B978-0-12-800953-6.00001-1

More theory from quantum mechanics and quantum information is necessary to understand how an unknown transformation can be learned. The key concepts are the Choi-Jamiołkowski duality (Section 13.1) that allows ancilla-assisted quantum process tomography (Section 13.2), compact Lie groups (Section 13.3) and their Clebsch-Gordan decomposition that define which representations are of interest, and covariant POVMs that will provide the optimal measurement (Section 13.4). These concepts are briefly introduced in the following sections, before we discuss how learning is performed.

13.1 Channel-State Duality

Given the properties of quantum states and transformations, noticing the similarities is inevitable. There is, in fact, a correspondence between quantum channels and quantum states. The Choi-Jamiołkowski isomorphism establishes a connection between linear maps from Hilbert space $\mathcal{H}_1$ to Hilbert space $\mathcal{H}_2$ and operators in the tensor product space $\mathcal{H}_1 \otimes \mathcal{H}_2$.

A quantum channel is a completely positive, trace-preserving linear map

$$\Phi : \mathcal{L}(\mathcal{H}_1) \mapsto \mathcal{L}(\mathcal{H}_2). \tag{13.1}$$

Here $\mathcal{L}(\mathcal{H})$ is the space of linear operators on $\mathcal{H}$. The map Φ takes a density matrix acting on the system in the Hilbert space $\mathcal{H}_1$ to a density matrix acting on the system in the Hilbert space $\mathcal{H}_2$.

Since density matrices are positive, Φ must preserve positivity, hence the requirement for a positive map. Furthermore, if an ancilla of some finite dimension n is coupled to the system, then the induced map $I_n \otimes \Phi$, where I_n is the identity map on the ancilla, must be positive—that is, $I_n \otimes \Phi$ is positive for all n. Such maps are called completely positive.

The last constraint, the requirement to preserve the trace, derives from the density matrices having trace 1.

For example, the unitary time evolution of a system is a quantum channel. It maps states in the same Hilbert space, and it is trace-preserving because it is unitary. More generic quantum channels between two Hilbert spaces act as communication channels which transmit quantum information. In this regard, quantum channels generalize unitary transformations.

We define a matrix for a completely positive, trace-preserving map Φ in the following way:

$$\rho_\Phi = \sum_{i,j=1}^{n} |e_i\rangle\langle e_j| \otimes \Phi(|e_i\rangle\langle e_j|). \tag{13.2}$$

This is called the Choi matrix of Φ. By Choi's theorem on completely positive maps, Φ is completely positive if and only if ρ_Φ is positive (Choi, 1975; Jamiołkowski, 1972). The operator ρ_Φ is a density matrix, and therefore it is the state dual to the quantum channel Φ.

The duality between channels and states is thus the linear bijection

$$\Phi \to \rho_\Phi. \tag{13.3}$$

This map is known as the Choi-Jamiołkowski isomorphism. This duality is convenient, as instead of studying linear maps, we can study the associated linear operators, which underlines the analogue that these maps are the generalization of unitary operators.

13.2 Quantum Process Tomography

Quantum state tomography is a process of reconstructing the state for a source of a quantum system by measurements on the system. In quantum process tomography, the states are known, but the quantum dynamics of the system is unknown—the goal is to characterize the process by probing it with quantum states (Chuang and Nielsen, 1997). The dynamics of a quantum system is described by a completely positive linear map:

$$\mathcal{E}(\rho) = \sum_i A_i \rho A_i^\dagger, \tag{13.4}$$

where $\sum_i A_i^\dagger A_i = I$ ensures that the map preserves the trace. This is called Krauss's form.

Let $\{E_i\}$ be an orthogonal basis for $\mathcal{B}(\mathcal{H})$, the space of bounded linear operators on $\mathcal{H}$. The operators A_i are expanded in this base as $A_i = \sum_m a_{im} E_m$. Thus, we have

$$\mathcal{E}(\rho) = \sum_{m,n} \chi_{mn} E_m \rho E_n^\dagger, \tag{13.5}$$

where $\chi_{mn} = \sum_i a_{mi} a_{ni}^*$. The map χ completely characterizes the process $\mathcal{E}$ in this basis.

There are direct and indirect methods for characterization of a process, and they are optimal for different underlying systems (Mohseni et al., 2008). Indirect methods rely on probe states and use quantum state tomography on the output states. With direct methods, experimental outcomes directly reveal the underlying dynamic process; there is no need for quantum state tomography.

Among indirect approaches, ancilla-assisted quantum process tomography is of special interest (Altepeter et al., 2003; D'Ariano and Lo Presti, 2003; Leung, 2003). It uses the Choi-Jamiołkowski isomorphism (Section 13.1) to perform state tomography and reveal the dynamic process. Consider the correspondence

$$\rho_{\mathcal{E}} \equiv (\mathcal{E} \otimes I)(|\Phi^+\rangle\langle\Phi^+|), \tag{13.6}$$

where $|\Phi^+\rangle = \sum_{i=1}^{d} (1/\sqrt{d})|i\rangle \otimes |i\rangle$ is the maximally entangled state of the system under study and an ancilla of the same size. The information about the dynamics is imprinted on the final state: our goal reduces to characterizing $\rho_{\mathcal{E}}$.

Assume that the initial state of the system with the ancilla is $\rho_{AB} = \sum_{i,j} \rho_{ij} E_i^A \otimes E_j^B$, where $\{E_i^A\}$ and $\{E_j^B\}$ are the operator bases in the respective space of linear operators of $\mathcal{H}_A$ and $\mathcal{H}_B$. The output state after applying $\mathcal{E}$ is

$$\rho'_{AB} = (\mathcal{E}_A \otimes I_B)(\rho_{AB}) = \sum_{i,j,m,n} \rho_{ij}\chi_{mn}E_m^A E_i^A E_n^{A\dagger} \otimes E_j^B. \quad (13.7)$$

Substituting $\tilde{\alpha}_{kj} = \sum_{m,n,i} \chi_{mn}\rho_{ij}\alpha_k^{m,i,n}$, where $E_m^A E_i^A E_n^{A\dagger} = \sum_k \alpha_k^{m,i,n} E_k^A$, we get the simple formula

$$\rho'_{AB} = \sum_{kj} \tilde{\alpha}_{kj} E_k^A \otimes E_j^B. \quad (13.8)$$

Notice that the values of $\alpha_k^{m,i,n}$ depend only on the choice of the operator basis. The $\tilde{\alpha}$ values can be obtained by joint measurements of the observables $E_k^A \otimes E_j^B$. They are obtained as

$$\tilde{\alpha}_{kj} = \mathrm{tr}(\rho'_{AB} E_k^{A\dagger} \otimes E_j^{B\dagger}). \quad (13.9)$$

Through the values of $\tilde{\alpha}_{kj}$, we obtain χ_{mn}, thus characterizing the quantum process.

While many classes of input states are viable, a maximally entangled input states yields the lowest experimental error rate, as ρ has to be inverted (Mohseni et al., 2008).

13.3 Groups, Compact Lie Groups, and the Unitary Group

Compact groups are natural extensions of finite groups, and many properties of finite groups carry over. Of them, compact Lie groups are the best understood. We will need their properties later, so we introduce a series of definitions that are related to these structures.

A group $\mathbf{G}$ is a finite or infinite set, together with an operation $\cdot$ such that

$$\begin{aligned} g_1 \cdot g_2 \in \mathbf{G} \quad & \forall g_1, g_2 \in \mathbf{G}, \\ (g_1 \cdot g_2) \cdot g_3 = g_1 \cdot (g_2 \cdot g_3) \quad & \forall g_1, g_2, g_3 \in \mathbf{G}, \\ \exists e \in \mathbf{G}: \quad e \cdot g = g \cdot e = g \quad & \forall g \in \mathbf{G}, \\ \forall g \in \mathbf{G} \;\; \exists h: \quad & g \cdot h = h \cdot g = e. \end{aligned} \quad (13.10)$$

We often omit the symbol $\cdot$ for the operation, and simply write $g_1 g_2$ to represent the composition. The element e is the unit element of the group. The element h in this context is the inverse of g, and it is often denoted as g^{-1}.

A topological group is a group $\mathbf{G}$ where the underlying set is also a topological space such that the group's operation and the group's inverse function are continuous functions with respect to the topology. The topological space of the group defines sets of neighborhoods for each group element that satisfy a set of axioms relating the elements and the neighborhoods. The neighborhoods are defined as subsets of the set of $\mathbf{G}$, called open sets, satisfying these conditions.

A function f between topological spaces is called continuous if for all $g \in \mathbf{G}$ and all neighborhoods N of $f(g)$ there is a neighborhood M of g such that $f(M) \subseteq N$. A special class of continuous functions is called homeomorphisms: these are continuous bijections for which the inverse function is also continuous. Topological groups introduce a sense of duality: we can perform the group operations on the elements of the set, and we can talk about continuous functions due to the topology.

A topological manifold is a topological space that is furthered characterized by having a structure which locally resembles real n-dimensional space. A topological space X is called locally Euclidean if there is a nonnegative integer n such that every point in X has a neighborhood which is homeomorphic to the Euclidean space $\mathbb{R}^n$. A topological manifold is a locally Euclidean Hausdorff space—that is, distinct points have distinct neighborhoods in the space.

To do calculus, we need a further definition. A differentiable manifold is a topological manifold equipped with an equivalence class of atlases whose transition maps are all differentiable. Here an atlas is a collection of charts, where each chart is a linear space where the usual rules of calculus apply. The differentiable transitions between the charts ensure there is a global structure. A smooth manifold is a differentiable manifold for which all the transition maps have derivatives of all orders—that is, they are smooth.

The underlying set of a Lie group is also a finite-dimensional smooth manifold, and in which the group operations of multiplication and inversion are also smooth maps. Smoothness of the group multiplication means that it is a smooth mapping of the product manifold $\mathbf{G} \times \mathbf{G}$ to $\mathbf{G}$.

A compact group is a topological group whose topology is compact. The intuitive view of compactness is that it generalizes closed and bounded subsets of Euclidean spaces. Formally, a topological space X is called compact if each of its open covers has a finite subcover—that is, for every collection $\{U_\alpha\}_{\alpha \in A}$ of open subsets of X such that $X = \bigcup_{\alpha \in A} U_\alpha$ there is a finite subset J of A such that $X = \bigcup_{i \in J} U_i$.

A compact Lie group has all the properties described so far, and it is a well-understood structure. An intuitive, albeit somewhat rough way to think about compact Lie groups is that they contain symmetries that form a bounded set.

To gain insight into these new definitions, we consider an example, the circle group, denoted by $\mathbf{U}(1)$, which is a one-dimensional compact Lie group. It is the unit circle on the complex plain with complex multiplication as the group operation:

$$\mathbf{U}(1) = \{z \in \mathbb{C} : |z| = 1\}. \tag{13.11}$$

The notation $\mathbf{U}(1)$ refers to the interpretation that this group can also be viewed as 1×1 unitary matrices acting on the complex plane by rotation about the origin.

Complex multiplication and inversion are continuous functions on this set; hence, it is a topological group.

Furthermore, the circle is a one-dimensional topological manifold. As multiplication and inversion are analytic maps on the circle, it is a smooth manifold. The unit circle is a closed subset of the complex plane; hence, it is a compact group. The circle group is indeed a compact Lie group.

If we generalize this example further, the unitary group $\mathbf{U}(n)$ of degree n is the group of $n \times n$ unitary matrices, with the matrix multiplication as the group operation. It is a finite-dimensional compact Lie group.

13.4 Representation Theory

Representation theory represents elements of groups as linear transformations of vector spaces, as these latter structures are much easier to understand and work with.

Let $\mathbf{G}$ be a group with a unit element e, and let $\mathcal{H}$ be a Hilbert space. A unitary representation of $\mathbf{G}$ is a function $U : \mathbf{G} \mapsto \mathcal{B}(\mathcal{H})$, $g \mapsto U_g$, where $\{U_g\}$ are unitary operators such that

$$U_{g_1 g_2} = U_{g_1} U_{g_2} \quad \forall g_1, g_2 \in \mathbf{G}, \qquad U_e = I. \tag{13.12}$$

Naturally, the unitaries themselves form a group; hence, if the map is a bijection, then $\{U_g\}$ is isomorphic to $\mathbf{G}$.

Recall that the Schrödinger evolution of a density matrix is given as $U\rho U^\dagger$ (see Equation 3.42). If the unitaries are from a representation of a group, we can study the automorphisms as the action of the group on the corresponding Hilbert space:

$$\mathcal{A}_g(\rho) = U_g \rho U_g^\dagger. \tag{13.13}$$

Following the properties of a unitary representation in Equation 13.12, we have

$$\mathcal{A}_{g_1 g_2}(\rho) = \mathcal{A}_{g_1}(\mathcal{A}_{g2}(\rho)), \quad \mathcal{A}_e(\rho) = \rho, \quad \mathcal{A}_{g^{-1}}(\mathcal{A}_g(\rho)) = \rho. \tag{13.14}$$

As we are working with quantum channels, the linear operators may not be unitary, but we restrict our attention to unitary channels.

Since the unitary representation is a group itself, this defines a projective representation, which is a linear representation up to scalar transformations

$$U_{g_1} U_{g_2} = \omega(g_1, g_2) U_{g_1 g_2} \quad \forall g_1, g_2 \in \mathbf{G}, \qquad U_e = I, \tag{13.15}$$

where $\omega(g_1, g_2)$ is a phase vector—that is, $|\omega(g_1, g_2)| = 1$. It is called a cocycle. We further require that

$$\omega(g_1, g_2 g_3)\omega(g_2, g_3) = \omega(g_1, g_2)\omega(g_1 g_2, g_3) \quad \forall g_1, g_2, g_3 \in \mathbf{G},$$
$$\omega(e, g) = \omega(g, e) = 1 \quad \forall g \in \mathbf{G}. \tag{13.16}$$

A representation of the unitary group is irreducible in an invariant subspace if the subspace does not have a proper subspace that is invariant. Any unitary representation $\{U_g\}$ of a compact Lie group can be decomposed into the direct sum of a discrete number of irreducible representations. This decomposition is essential in studying the properties of the representation.

More formally, let $\mathcal{H}$ be a Hilbert space and let $\{U_g | g \in \mathbf{G}\}$ be a projective representation. An invariant subspace $W \subseteq \mathcal{H}$ is a subspace such that $U_g W = W$ $\forall g \in \mathbf{G}$. Then, the representation $\{U_g\}$ is *irreducible* in W if there is no proper invariant subspace V, $\{0\} \neq V \subset W$.

If $\mathbf{G}$ is a compact Lie group, and $\{U_g | g \in \mathbf{G}\}$ a projective representation of G on the Hilbert space $\mathcal{H}$, then U_g can be decomposed into the direct sum of a discrete number of irreducible representations.

Two projective irreducible representations $\{U_g\}$ and $\{V_h\}$ acting in the Hilbert spaces $\mathcal{H}_1$ and $\mathcal{H}_2$, respectively, are called equivalent if there is an isomorphism $T : \mathcal{H}_1 \mapsto \mathcal{H}_2$ such that $T^\dagger T = I_{\mathcal{H}_1}, TT^\dagger = I_{\mathcal{H}_2}$, and $TU_g = V_g T \; \forall g \in \mathbf{G}$. In other words, $V_g = TU_g T^\dagger$, which implies that the two representations have the same cocycle. The isomorphism T is called an intertwiner.

Let $\{U_g\}$ and $\{V_g\}$ be two equivalent projective irreducible representations. Then, any operator $O : \mathcal{H}_1 \mapsto \mathcal{H}_2$ such that $OU_g = V_g O \; \forall g \in \mathbf{G}$ has the form $O = \lambda T$, where $\lambda \in \mathbb{C}$ is a constant, and $T : \mathcal{H}_1 \mapsto \mathcal{H}_2$ is an intertwiner.

In matrix form, this means that if a matrix O commutes with all matrices in $\{U_g\}$, then O is a scalar matrix. This is known as the Shur lemma.

If the two projective irreducible representations $\{U_g\}$ and $\{V_g\}$ are inequivalent, then the only operator $O : \mathcal{H}_1 \mapsto \mathcal{H}_2$ such that $OU_g = V_g O \; \forall g \in \mathbf{G}$ is the null operator $O = 0$.

We decompose a group representation $\{U_g\}$ acting in the Hilbert space $\mathcal{H}$ in a discrete number of irreducible representations, each of them acting on a different subspace:

$$U_g = \oplus_{\mu \in S} \oplus_{i=1}^{m_\mu} U_g^{\mu,i}, \tag{13.17}$$

where S is the set of equivalence classes of irreducible representations contained in the decomposition, and m_μ is the multiplicity of equivalent irreducible representations in the same equivalence class. The irreducible representations are naturally associated with irreducible orthogonal subspace of the Hilbert space $\mathcal{H}$:

$$\mathcal{H} = \oplus_{\mu \in S} \oplus_{i=1}^{m_\mu} \mathcal{H}_i^\mu. \tag{13.18}$$

With use of the Shur lemma, it is trivial to show that if $O \in \mathcal{B}(\mathcal{H})$ is in the commutator of $\{U_g\}$ ($[O, U_g] = 0 \; \forall g \in G$), then O can be written as

$$O = \oplus_{\mu \in S} \oplus_{i=1}^{m_\mu} \lambda_{ij}^\mu T_{ij}^\mu, \tag{13.19}$$

where $\lambda_{ij}^\mu \in \mathbb{C}$ are suitable constants.

We simplify the decomposition of the Hilbert space further by replacing a direct sum with a tensor product of an appropriate complex space. Consider the invariant subspaces $\mathcal{H}_i^\mu$ and $\mathcal{H}_i^\mu$, which correspond to equivalent irreducible representations in the same equivalence class, and two orthonormal bases in theses subspaces $B_i^\mu = \{|\mu, i, n\rangle | n = 1, \ldots, d_\mu\}$ and $B_j^\mu = \{|\mu, j, n\rangle | n = 1, \ldots, d_\mu\}$ such that

$$|\mu, i, n\rangle = T_{ij}^\mu |\mu, j, n\rangle \quad \forall n = 1, \ldots, d_\mu. \tag{13.20}$$

This view enables us to identify a basis vector with the following tensor product:

$$|\mu, i, n\rangle \cong |\mu, n\rangle \otimes |i\rangle. \tag{13.21}$$

Here $\{|\mu, i, n\rangle | n = 1, \ldots, d_\mu\}$ and $\{|i\rangle | i = 1, \ldots, m_\mu\}$ are orthonormal bases of $\mathcal{H}_\mu$ and $\mathbb{C}^{m_\mu}$, respectively. $\mathcal{H}_\mu$ is called the representation space, and $\mathbb{C}^{m_\mu}$ is the multiplicity space. With this identification, we obtain the Clebsch-Gordan tensor product structure—also called the Clebsch-Gordan decomposition:

$$\mathcal{H} = \oplus_{\mu \in S} \mathcal{H}_\mu \otimes \mathbb{C}^{m_\mu}. \tag{13.22}$$

The intertwiner in this notation reduces to the tensor product of an identity operator and a projection between complex spaces:

$$T^\mu_{ij} = I_{d_\mu} \otimes |i\rangle\langle j|. \tag{13.23}$$

The matching unitary representation $\{U_g\}$ has a simple form:

$$U_g = \oplus_{\mu \in S} U^\mu_g \otimes I_{m_\mu}, \tag{13.24}$$

Covariant POVMs play a major role in the optimal estimation of quantum states: they address the optimal extraction of information from families of states that are invariant under the action of a symmetry group. A POVM $P(d\omega)$ is covariant if and only for any state $\rho \in S(\mathcal{H})$ the probability distribution $p(B|\rho) = \text{tr}(P(B)\rho)$ is group-invariant—that is,

$$p(B|\rho) = p(gB|\mathcal{A}_g(\rho)) \quad \forall B \in \sigma(\omega), \tag{13.25}$$

where $gB = \{g\omega|\omega \in B\}$ and $\mathcal{A}_g(\rho) = U_g \rho U_g^\dagger$.

Let **G** be a Lie group. For any fixed group element $h \in \mathbf{G}$, the map $g \mapsto hg$ is a diffeomorphism, and transforms the region $B \subseteq G$ into the region $hB = \{hg|g \in B\}$. A measure $\mu_\text{L}(dg)$ on the Lie group **G** is called *left-invariant* if $\mu_\text{L}(gB) = \mu_\text{L}(B)$, for any group element $g \in \mathbf{G}$ and for any region $B \subseteq \mathbf{G}$. Right invariance is defined similarly. While any Lie group has a left-invariant and a right-invariant measure, they do not necessarily coincide. If they do, the group is called unimodular, and the measure is called invariant.

Let G be a locally compact unimodular group, and let $\{U_g\}$ be a projective representation of **G** in the Hilbert space $\mathcal{H}$. Let $H_0 \subseteq \mathbf{G}$ be a compact subgroup. Then, a POVM $P(d\omega)$ with outcome space $\Omega = G/H_0$ is covariant if and only if it has the form

$$P(\text{d}\omega) = U_g(\omega)\Xi U_g(\omega)^\dagger \nu(\text{d}\omega), \tag{13.26}$$

where $g(\omega) \in \mathbf{G}$ is any representative element of the equivalence class $\omega \in \Omega$, $\nu(\text{d}\omega)$ is the group-invariant measure over ω, and Ξ is an operator satisfying the properties

$$\begin{aligned} &\Xi \geq 0 \\ &[\Xi, U_h] = 0 \quad \forall h \in H_0 \\ &\int_\mathbf{G} U_g \Xi U_g^\dagger \text{d}g = I, \end{aligned} \tag{13.27}$$

where dg is the invariant Haar measure over the group **G**.

This establishes a bijective correspondence between a covariant POVM with outcome space $\Omega = G/H_0$ with a single operator Ξ, which is called the seed of the POVM.

Given a Clebsch-Gordan tensor product structure for the elements of a representation $\{U_g\}$ in the form of Equation 13.24, the normalization condition in the theorem for the seed of the POVM can be written in the simple form

$$\mathrm{tr}_{H_\mu}(\Xi) = d_\mu I_{m_\mu}. \tag{13.28}$$

13.5 Parallel Application and Storage of the Unitary

The task is simple: we have a black box that implements an unknown unitary U and we can make N calls to it to identify the unitary. This is fundamentally different from the classical setting in how the training instances are acquired. In regression, we have labeled examples $\{(\mathbf{x}_1, y_1), \ldots, (\mathbf{x}_N, y_N)\}$ to which we fit a function. In the quantum setting, we assume that we are able to use the function generating the labels N times. While in the classical scenario there is little difference between using the pairs $(\mathbf{x}_i, y_i)$, or $(\mathbf{x}_i, f(\mathbf{x}_i))$, where f is the original function generating the process, in the quantum setting, we must have access to U.

Regression in a classical learning setting finds an estimator function in a family of functions characterized by some parameter θ (Sections 8.1 and 8.2). In the quantum learning setting, the family of functions is the unitary representation $\{U_g\}$ of a compact Lie group $\mathbf{G}$, and the parameter we are estimating is the group element $g \in \mathbf{G}$. We omit the index of U_g in the rest of the discussion. The generic process of learning and applying the learned function to a single new output is outlined in Figure 13.1.

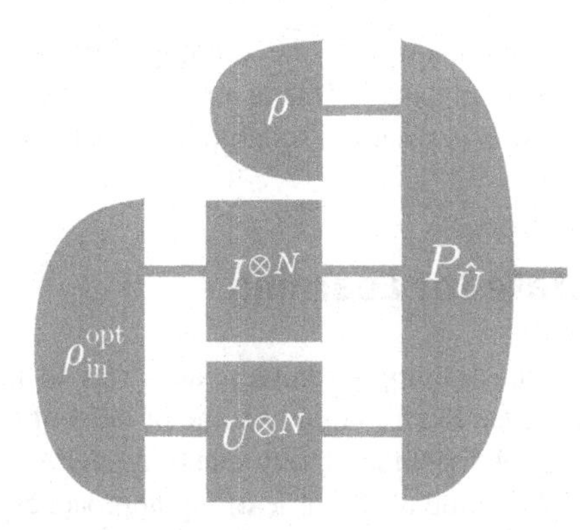

Figure 13.1 Outline of quantum learning of a unitary transformation U. An optimal input state $\rho_{\text{in}}^{\text{opt}}$ is a maximally entangled state with an ancilla, consisting of a direct sum of identity operators over irreducible spaces of $U^{\otimes N}$. This state is acted on by N parallel calls of U—that is, by $U^{\otimes N}$—while the action on the ancilla is trivial ($I^{\otimes N}$). To apply the learned unitary $\hat{U}$, we perform an optimal POVM measurement strategy $P_{\hat{U}}$ on the output state, and apply the estimated unitary on the new state ρ.

U acts on a finite d-dimensional Hilbert space, and performs a deterministic transformation belonging to a given unitary representation of a compact Lie group (Chiribella et al., 2005). If we are guaranteed to have N calls of U, two immediate problems arise:

1. How do we store the approximated unitary?
2. How do we dispose the N uses, in parallel or in sequence?

The first question is easier to address: the Choi-Jamiołkowsky duality in Equation 13.3 enables us to treat the transformation of the quantum channel as a state.

To find out the answer to the second question, we define our objective: we maximize the fidelity of the output state with the target state, averaging over all pure input states and all unknown unitaries in the representation of the group with an invariant measure. Thus, the objective function, the channel fidelity is written as

$$F = \frac{1}{d^2} \int_{\mathrm{G}} \operatorname{tr} \left\{ L \left[|U\rangle\langle U \otimes \left(|U\rangle\langle U|^{\otimes N} \right)^{\top} \right] \right\} \mathrm{d}U, \tag{13.29}$$

where we work in the computational basis. Here L is a positive operator describing the operation of the quantum channel, *including* the application of U to a new input state ρ that is not among the training instances. This inclusion of the new input in the objective already hints at transduction.

If we take the fidelity of output quantum states as the figure of merit, the optimal storage is achieved by a parallel application of the unitaries on an input state. Denote by $\mathcal{H}_i$ the Hilbert space of all inputs of the N examples, and denote by $\mathcal{H}_o$ the Hilbert space of all outputs. With $U^{\otimes N}$ acting on $\mathcal{H}_i$, we have the following (Bisio et al., 2010): the optimal storage of U can be achieved by applying $U^{\otimes N} \otimes I^{\otimes N}$ on a suitable multipartite input state $|\phi\rangle \in \mathcal{H}_o \otimes \mathcal{H}_i$.

The corollary of this lemma is that the training time is constant: it does not depend on the number of examples N. This is a remarkable departure from classical algorithms. The next question is what the suitable input state might be.

13.6 Optimal State for Learning

Classical learning takes the training examples, and tries to make the best use of them. Certain algorithms even learn to ignore some or most of the training examples. For instance, support vector machines build sparse models using a fraction of the training data (Section 7.2). Some classical learning approaches may ask for specific extra examples, as, for instance, active learning does. Active learning is a variant of semisupervised learning in which the learning algorithm is able to solicit labels for problematic unlabeled instances from an appropriate information source—for instance, from a human annotator (Section 2.3). Similarly to the semisupervised setting, there are some labels available, but most of the examples are unlabeled. The task in a learning iteration is to choose the optimal set of unlabeled examples for which the algorithm solicits labels.

Optimal quantum learning of unitaries resembles active learning in a sense: it requires an optimal input state. Since the learner has access to U, by calling the transformation on an optimal input state, the learner ensures that the most important characteristics are imprinted on the approximation state of the unitary.

The quantum learning process is active, but avoids induction. In inductive learning, based on a set of data points—labeled or unlabeled—we infer a function that will be applied to unseen data points. Transduction avoids this inference to the more general, and it infers from the particular to the particular (Section 2.3). This way, transduction asks for less: an inductive function implies a transductive one. That is, given our training set of points $\{(\mathbf{x}_1, y_1), \ldots, (\mathbf{x}_N, y_N)\}$, we are trying to label just M more points $\{\mathbf{x}_{N+1}, \ldots, \mathbf{x}_{N+M}\}$. Inductive learning solves a more general problem of finding a rule to label any future object.

Transduction resembles instance-based learning, a family of algorithms that compare new problem instances with training instances—K-means clustering and K-nearest neighbors are two examples (Sections 5.3 and 7.1). If some labels are available, transductive learning resembles semisupervised learning. Yet, transduction is different: instance-based learning can be inductive, and semisupervised learning is inductive, whereas transductive learning avoids inductive reasoning by definition. In a way, the goal in transductive learning is actually to minimize the test error, instead of the more abstract goal of maximizing generalization performance (El-Yaniv and Pechyony, 2007).

The way transduction manifests itself in the optimal learning of a unitary transform is also via the input state. The input state is a superposition, and the probabilities of the superposition depend on M.

A further requirement of the input state is that it has to be maximally entangled. To give an example of why entanglement is necessary, consider superdense coding: two classical bits are sent over a quantum channel by a single qubit. The two communicating parties, Alice and Bob, each have half of two entangled qubits—a Bell state. Alice applies one of four unitary transformations on her part, translating the Bell state into a different one, and sends the qubit over. Bob measures the state, and deduces which one of the four operations was used, thus retrieving two classical bits of information. The use of entanglement with an ancillary system improves the discrimination of unknown transformations; this motivates the use of such states in generic process tomography (Chiribella et al., 2005).

To bring active learning, transduction, and entanglement into a common framework, we use the Clebsch-Gordon decomposition of $U^{\otimes N}$. The following lemma identifies the optimal input state (Bisio et al., 2010). The optimal input state for storage can be taken to be of the form

$$|\phi\rangle = \oplus_{j\in \mathrm{Irr}(U^{\otimes N})}\sqrt{\frac{p_j}{d_j}}|I_j\rangle \in \tilde{\mathcal{H}}, \tag{13.30}$$

where p_j are probabilities, the index j runs over the set $\mathrm{Irr}(U^{\otimes N})$ of all irreducible representations $\{U_j\}$ contained in the decomposition of $\{U^{\otimes N}\}$, d_j is the dimension of the corresponding subspace $\mathcal{H}_j$, and $\tilde{\mathcal{H}} = \oplus_{j\in \mathrm{Irr}(U^{\otimes N})}(\mathcal{H}_j \otimes \mathcal{H}_j)$ is a subspace of

$\mathcal{H}_o \otimes \mathcal{H}_i$ carrying the representation $\tilde{U} = \bigoplus_{j\in\mathrm{Irr}(U^{\otimes N})}(U_j \otimes I_j)$, I_j being the identity in $\mathcal{H}_j$.

The optimal state for the estimation of an unknown unitary is always a superposition of maximally entangled states, similarly to the case of quantum process tomography (Mohseni et al., 2008). Yet, this result is subtler: what matters here is the entanglement between a space in which the action of the group is irreducible and a space in which the action of the group is trivial (Chiribella et al., 2005).

Requiring this form of input state is a special variant of active learning in which we are able to request the unitary transformation of this maximally entangled state in parallel.

From the optimal input state in Equation 13.30, it is not immediately obvious why learning the unitary is a form of transduction. The way the entangled state is constructed is through irreducible representations, which in turn encode symmetries and invariances of U. These characteristics depend primarily on how U should act on the ρ_{in} states to gain the target ρ_{out} states. The probabilities p_j, however, depend on M, and thus on the test data, which indeed confirms that this learning scenario is transduction. The retrieval strategy that applies U performs measurements, and to obtain the highest fidelity, p_j must be tuned. We detail this in Section 13.7.

Choosing the optimal input state is necessary in similar learning strategies, such as a binary classification scheme based on the quantum state tomography of state classes through Helstrom measurements (Guţă and Kotłowski, 2010). In this case, the POVM is the learned decision model, and it is probabilistic: the quantum system has an additional parameter for the identification of the training set that makes it easier to predict the output. Similarly, Bisio et al. (2011) require a maximally entangled state to learn measurements, and then perform transduction to a new state; this is the $M = 1$ case. Generalizing this method to larger M values is difficult.

13.7 Applying the Unitary and Finding the Parameter for the Input State

The intuitive appeal of the optimal input state in Equation 13.30 is that states in this form are easy to distinguish. Consider the inner product of two such states, $|\phi_g\rangle$ and $|\phi_h\rangle$, for the $N = 1$ case:

$$\langle\phi_g|\phi_h\rangle = \sum_{j\in\mathrm{Irr}(U)} \frac{p_j}{d_i}\mathrm{tr}(U^j_{g^{-1}h}). \tag{13.31}$$

If p_j is chosen such that $p_j = \lambda d_j$, then this equation becomes an approximation of the Dirac delta function in $\mathbf{G}$: $\delta_{gh} = \sum_{j\in\mathrm{Irr}(G,\omega)} \frac{p_j}{d_i}\mathrm{tr}(U^j_{g^{-1}h})$ (Chiribella, 2011). Thus the choice of the input state approximates optimal separation.

A measure-and-prepare strategy is optimal for applying the learned function an arbitrary number of times. This strategy consists in measuring the state $|\phi_U\rangle$ in the quantum memory with an optimal POVM, and performing the unitary on the new input state. Maximizing the fidelity with the measurement includes tuning p_j.

There is no loss of generality in restricting the search space to covariant POVMs (Holevo, 1982), which corresponds to the estimation of unitaries that are picked out from a representation at random. Covariant POVMs of the form $M(g) = U_g \Xi U_g^\dagger$ are of interest (see Equation 13.27), with a positive operator Ξ satisfying the normalizing condition $\int_G M(g)dg = I$ (Chiribella et al., 2005). The following theorem provides the optimal measurement to retrieve the approximated unitary (Bisio et al., 2010):

The optimal retrieval of U from the memory state $|\phi_U\rangle$ is achieved by measuring the ancilla with the optimal covariant POVM in the form $\Xi = |\eta\rangle\langle\eta|$ (see Equation 13.27)—namely,

$$P_{\hat{U}} = |\eta_{\hat{U}}\rangle\langle\eta_{\hat{U}}|, \tag{13.32}$$

given by

$$|\eta_{\hat{U}}\rangle = \oplus_j \sqrt{d_j} |\hat{U}_j\rangle, \tag{13.33}$$

and, conditionally on outcome $\hat{U}$, by performing the unitary $\hat{U}$ on the new input system.

Considering the fidelity of learning the unitary in Equation 13.29 with this covariant POVM, we can write the optimal probability coefficients as follows (Chiribella et al., 2005):

$$p_j = \frac{d_j m_j}{d^N}, \tag{13.34}$$

where m_j is the multiplicity of the corresponding space. This is true for the $M = 1$ case. The more generic case for arbitrary M takes the global fidelity between the estimate channel C_U and $U^{\otimes M}$—that is, the objective function averages over $(|\text{tr}(U^\dagger C_U)|/d)^2$. Hence, the exact values of p_j always depend on M, tying the optimal input state to the number of applications, making it a case of transduction.

Generally, fidelity scales as $F \propto \frac{1}{N^2}$, for instance, for qubit states. The measurement will be optimal in the fidelity of quantum states, and it is also optimal for the maximization of the single-copy fidelity of all local channels. The fidelity does not degrade with repeated applications. If we thus measure the unitary, we use an incoherent classical storage of the process—that is, we perform a procedure similar to ancilla-assisted quantum process tomography. This incoherent strategy detaches the unitary from the examples, and we store it in classical memory. While the optimal input state necessarily depends on M, the incoherent process induces a function that can be applied to any number of examples, which is characteristic to inductive learning. It is a remarkable crossover between the two approaches.

While this is not optimal, it is interesting to consider what happens if we do not measure U, but store it in quantum memory, and retrieve it from there when needed. Naturally, with every application, the state degrades. Yet, this situation is closer to pure transductive learning. The stored unitary transform depends solely on the data instances. The coherent strategy does not actually learn the process, but rather keeps the imprint in superposition.

From the perspective of machine learning, restricting the approximating function to unitaries is a serious constraint. It implies that the function is a bijection, ruling classification out. This method is relevant in regression problems. A similar approach that also avoids using quantum memory applies to binary classification: the unitary evolution is irrelevant in this case, and the optimality of measurement ensures robust classification based on balanced sets of examples (Sentís et al., 2012).

Boosting and Adiabatic Quantum Computing

14

Simulated annealing is an optimization heuristic that seeks the global optimum of an objective function. The quantum flavor of this heuristic is known as quantum annealing, which is directly implemented by adiabatic quantum computing (Section 14.1). The optimization process through quantum annealing solves an Ising model, which we interpret as a form of quadratic unconstrained binary optimization (Sections 14.2 and 14.3).

Boosting—that is, combining several weak classifiers to build a strong classifier—translates to a quadratic unconstrained binary optimization problem. In this form, boosting naturally fits adiabatic quantum computing, bypassing the circuit model of quantum computing (Section 14.4).

The adiabatic paradigm offers reduced computational time and improved generalization performance owing to the nonconvex objective functions favored under the model (Section 14.5). The adiabatic paradigm, however, suffers from certain challenges; the following are the most important:

- Weights map to qubits, resulting in a limited bit depth for manipulating the weights of optimization problems.
- Interactions in an Ising model are considered only among at most two qubits, restricting optimization problems to maximum second-order elements.

Contrary to expectations, these challenges improve sparsity and thus affect generalization performance (Section 14.6).

Adiabatic quantum computing has a distinct advantage over competing quantum computing methods: it has already been experimentally demonstrated in multiple-qubit workloads. The hardware implementation is controversial for many reasons—for instance, entanglement between qubits, if it exists, belongs to a category that is easily simulated on classical systems. Therefore, it is good practice to keep the theory of adiabatic quantum computing and its current practice separate. The current implementations, nevertheless, impose further restrictions, which we must address:

- Not all pairs of qubits are connected owing to engineering constraints: there is a limited connectivity between qubits, albeit the connections are known in advance.
- Optimization happens at a temperature slightly higher than absolute zero; hence, there is thermal noise and excited states may be observed.

Quantum Machine Learning. http://dx.doi.org/10.1016/B978-0-12-800953-6.00001-1

We must adjust the objective function and the optimization process to accommodate these constraints (Section 14.7).

Addressing all these constraints means multiple additional optimization steps using a classical computer: we must optimize the free parameter in quantum boosting, the free parameter in the bit width, and the mapping to actual qubit connectivity. If we discount the complexity of classical optimization steps, the size of the adiabatic gap controls the execution time, which may be quadratically faster than in classical boosting (Section 14.8).

14.1 Quantum Annealing

Global optimization problems are often NP-hard. In place of exhaustive searches, heuristic algorithms approximate the global optimum, aiming to find a satisfactory solution. A metaheuristic is a high-level framework that allows the generation of heuristic algorithms for specific optimization problems. Simulated annealing is a widely successful metaheuristic (Sörensen, 2013).

As in the case of many other heuristic methods, simulated annealing borrows a metaphor from a physical process. In metallurgy, annealing is a process to create a defect-free crystalline solid. The process involves heating the material—possibly only locally—then cooling it in a controlled manner. Internal stress is relieved as the rate of diffusion of the atoms increases with temperature, and imperfectly placed atoms can attain their optimal location with the lowest energy in the crystal structure.

Simulated annealing uses the metaphor of this thermodynamic process by mapping solutions of an optimization problem to atomic configurations. The random dislocations in annealing change the value of the solution. Simulated annealing avoids getting trapped in a local minimum by accepting a solution that does not lower the value with a certain probability. Accepting worse solutions was, in fact, the novelty of this metaheuristic. The probability is controlled by the "temperature": the higher it is, the more random movements are allowed. The temperature decreases as the heuristic explores the search space.

This technique is often used when the search space is discrete. Neighboring states in the search space are thus permutations of the discrete variables. Simulated annealing converges to the global optimum under mild conditions, but convergence can be slow (Laarhoven and Aarts, 1987).

Quantum annealing replaces the thermodynamic metaphor, and uses the tunneling field strength in lieu of temperature to control acceptance probabilities (Figure 14.1) (Finnila et al., 1994)). Neighborhood selection in simulated annealing does not change throughout the simulation, whereas tunneling field strength is related to kinetic energy, and consequently the neighborhood radius is decreased in subsequent iterations. Its further advantage is that the procedure maps to an actual physical process. Hence, it is possible to implement optimization directly on quantum hardware, bypassing the higher-level models of quantum computing.

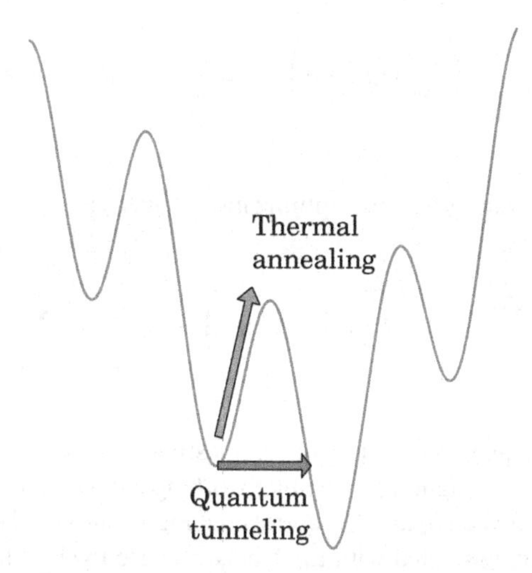

Figure 14.1 Comparison of thermal and quantum annealing: thermal annealing often has a higher barrier than quantum tunneling.

Quantum annealing, however, is prone to getting trapped in metastable exited states, and may spend a long time in these states, instead of reaching the ground-state energy. The quantum adiabatic theorem helps solve this problem.

14.2 Quadratic Unconstrained Binary Optimization

We consider an alternative formulation of boosting. Instead of an exponential loss function in Equation 9.4, we define a regularized quadratic problem to find the optimal weights. The generic problem of quadratic unconstrained binary optimization (QUBO) is defined as follows:

$$\operatorname{argmin} \sum_{i,j=1}^{N} w_{ij} x_i x_j, \tag{14.1}$$

such that

$$x_i \in \{0, 1\} \qquad i = 1, \ldots, N. \tag{14.2}$$

Given a set of weak learners $\{h_s | h_s : \mathbb{R}^d \mapsto \{-1, 1\}, s = 1, 2, \ldots, K\}$, boosting translates to the following objective function (Neven et al., 2008):

$$\underset{\mathbf{w}}{\operatorname{argmin}} \left(\frac{1}{N} \sum_{i=1}^{N} \left(\sum_{s=1}^{K} w_s h_s(\mathbf{x}_i) - y_i \right)^2 + \lambda \|\mathbf{w}\|_0 \right)$$

$$= \underset{\mathbf{w}}{\operatorname{argmin}} \left(\frac{1}{N} \sum_{i=1}^{N} \left[\left(\sum_{s=1}^{K} w_s h_s(\mathbf{x}_i) \right)^2 - 2 \sum_{s=1}^{K} w_s h_s(\mathbf{x}_i) y_s + y_i^2 \right] + \lambda \|\mathbf{w}\|_0 \right). \tag{14.3}$$

Since y_s^2 is just a constant offset, the optimization reduces to

$$\underset{\mathbf{w}}{\operatorname{argmin}} \left(\frac{1}{N} \sum_{s=1}^{K} \sum_{t=1}^{K} w_s w_t \left(\sum_{i=1}^{N} h_s(\mathbf{x}_i) h_t(\mathbf{x}_i) \right) - \frac{2}{N} \sum_{s=1}^{K} w_s \sum_{i=1}^{N} h_s(\mathbf{x}_i) y_i + \lambda \|\mathbf{w}\|_0 \right). \tag{14.4}$$

If we discretize this equation (Section 14.6), boosting becomes a QUBO.

Equation 14.4 offers insight into the inner workings of the optimization. It depends on the correlations between h_i and the corresponding y_i outputs. Those weak learners h_i that have an output correlated with the labels y_i cause the bias term to be lowered; thus, the corresponding w_i weights have a higher probability of being 1.

The quadratic terms, which express coupling, depend on the correlation between the weak classifiers. Strong correlations will force the corresponding weights to be lowered; hence, these weights have a higher probability of being 0.

QUBO is an NP-hard problem, yet, since the search space of QUBO is discrete, it is well suited for simulated annealing (Katayama and Narihisa, 2001) and quantum annealing (Neven et al., 2008). Solutions on classical computers are simulated by tabu search, which is based on local searches of a neighborhood (Glover, 1989; Palubeckis, 2004).

14.3 Ising Model

The Ising model is a model in statistical mechanics to study the magnetic dipole moments of atomic spins. The spins are arranged in a lattice, and only neighboring spins are assumed to interact. The optimal configuration is given by

$$\underset{\mathbf{s}}{\operatorname{argmin}}(\mathbf{s}^\top J \mathbf{s} + \mathbf{h}^\top \mathbf{s}), \tag{14.5}$$

where $s_i \in \{-1, +1\}$—these variables are called spins. The J operator describes the interactions between the spins, and $\mathbf{h}$ stands for the impact of an external magnetic field. The Ising model transforms to a QUBO with a change of variables $\mathbf{s} = 2\mathbf{w} - 1$.

The Ising objective function is represented by the following Hamiltonian:

$$H_{\mathrm{I}} = \sum_{i,j} J_{i,j} \sigma_i^z \sigma_j^z + \sum_i h_i \sigma_i^z, \tag{14.6}$$

where σ_i^z is the Pauli Z operator acting on qubit i.

The ground-state energy of this Hamiltonian will correspond to the optimum of the QUBO. Thus, we are seeking the following ground state:

$$\underset{\mathbf{s}}{\operatorname{argmin}}(\mathbf{s}^{\top} H_{\mathrm{I}} \mathbf{s}). \tag{14.7}$$

Take the base Hamiltonian of an adiabatic process as

$$H_{\mathrm{B}} = \sum_{i} \left(\frac{1 - \sigma_i^x}{2} \right). \tag{14.8}$$

Inserting H_{B} and H_{I} in Equation 3.56, we have an implementable formulation of the Ising problem on an adiabatic quantum processor:

$$H(\lambda) = (1 - \lambda)H_{\mathrm{B}} + \lambda H_{\mathrm{I}}, \tag{14.9}$$

14.4 QBoost

The Ising model allows mapping the solution of a QUBO problem to an adiabatic process. With this mapping, we are able to define QBoost, a boosting algorithm that finds the optimum of a QUBO via adiabatic optimization.

Denote the number of weak learners by K; this equals N if no other constraints apply (see Section 14.7 for modifications if N is too large to be handled by the hardware). Further, let W denote the set of weak learners in the final classifier, and let T denote its cardinality. Algorithm 7 describes the steps of QBoost.

As in LPBoost (Section 9.3), the feature space in QBoost is the output of weak learners. Hence, this method is not constrained by sequentiality either, unlike in the case of adaptive boosting (AdaBoost).

14.5 Nonconvexity

Nonconvexity manifests itself in two forms. One is the regularization term, the cardinality of the set of weights $\{w_i\}$. The other source of nonconvexity is the loss function, albeit it is usually taken as a quadratic function.

The regularization term in classical boosting is either absent or a convex function. Convexity yields analytic evidence for global convergence, and it often ensures the feasibility of a polynomial time algorithm. Nonconvex objective functions, on the other hand, imply an exhaustive search in the space. Adiabatic quantum computing overcomes this problem by the approximating Hamiltonian: we are not constrained by a convex regularization term.

In the basic formulation of QBoost, we used a quadratic loss function, but we may look into finding a nonconvex loss function that maps to the quantum hardware. If there is such a function, its global optimum will be approximated efficiently by an adiabatic process—as in the case of a quadratic loss function.

ALGORITHM 7 QBoost

```
Require: Training and validation data, dictionary of weak
  classifiers
Ensure: Strong classifier
  Initialize weight distribution d over training samples as uniform
  distribution ∀i : d(i) = 1/N.
  Set T ← 0 and W ← ∅
  repeat
    Select the set 𝒲 of K − T weak classifiers h_i from the
    dictionary that have the smallest weighted training error rates
    for λ = λ_min to λ_max do
      Find w*(λ) = argmin_w(∑_{i=1}^N (∑_{h_t∈W∪𝒲} w_t h_t(x_i)/K − y_i)^2 + λ‖w‖_0)
      Set T(λ) ← ‖w*(λ)‖_0
      Construct H(x; λ) ← sign(∑_{t=1}^T w*_t(λ)h_t(x))
      Measure validation error E_val(λ) of H(x; λ) on unweighted
      validation set
    end for
    λ* ← argmin_λ E_val(λ)
    Update T ← T(λ*) and W ← {h_t|w*_t(λ*) = 1}
    Update weights d(i) ← d(i)(∑_{h_t∈W} ht(x)/T − y_i)^2
    Normalize d(i) ← d(i)/∑_{i=1}^N d(i)
  until validation error E_val stops decreasing
```

First, if we replace the quadratic loss function in Equation 14.3 by the optimal 0-1 loss function, analytic approaches no longer apply in solving the optimization problem; in fact, it becomes NP-hard (Feldman et al., 2012). Convex approximation circumvents this problem, but label noise adversely affects convex loss functions (Section 9.3).

To approximate the 0-1 loss function with a QUBO model, we are seeking a loss function that is a quadratic function. The simple quadratic loss in Equation 14.3 is a convex variant. To make this loss function robust to label noise, we modify it with a parameterization. We define q-loss as

$$L_q(m) = \min\left((1-q)^2, (\max(0, 1-m))^2\right), \tag{14.10}$$

where $q \in (-\infty, 0]$ is a parameter (Denchev et al., 2012). This loss function is no longer convex, but it does not map to the QUBO model. Instead of calculating q-loss directly, we approximate it via a family of quadratic functions characterized by a variational parameter $t \in \mathbb{R}$. The following is equivalent to Equation 14.10:

$$L_q(m) = \min_t \left((m-t)^2 + (1-q)^2 \frac{1-\operatorname{sign}(t-1)}{2}\right). \tag{14.11}$$

Since this equation includes finding a minimum, the objective function becomes a joint optimization, as it has to find the optimum for both **w** and t. The advantage is that it directly maps to a QUBO; hence, it can be solved efficiently by adiabatic quantum computing, albeit the actual formulation is tedious. The following formula assumes a Euclidean regularization term on linear decision stumps, with q-loss as the loss function:

$$\underset{\mathbf{w},b,\mathbf{t}}{\operatorname{argmin}} \sum_{s,t=1}^{K} w_s w_t \left[\frac{1}{N}\sum_{i=1}^{N} x_{is}x_{it}\right] + b^2[1] + b\sum_{s=1}^{K} w_s \left[\frac{2}{N}\sum_{i=1}^{N} x_{is}\right]$$
$$+\sum_{s=1}^{K}\sum_{i=1}^{N} w_s t_i \left[\frac{-2y_i x_{is}}{N}\right] + \sum_{i=1}^{N}\left\{t_i^2\left[\frac{1}{N}\right] + bt_i\left[\frac{-2y_i}{N}\right] + t_{id_t}\left[\frac{-(1-q)^2}{N}\right]\right\}$$
$$+\sum_{s=1}^{K} w_s^2[\lambda], \tag{14.12}$$

where t_{id_t} is the binary indicator bit of $\operatorname{sign}(t_i - 1)$. The values in square brackets are the entries in the constraint matrix in the QUBO formulation. This formulation assumes continuous variables, so it needs to be discretized to map the problem to qubits.

Finding the optimal t reveals how q-loss handles label noise. In the second formulation of q-loss (Equation 14.11), t can change sign for a large negative margin, thus flagging mislabeled training instances. For any value of m, the optimal t of the minimizer (Equation 14.11) belongs to one of three classes:

1. $m \geq 1 \Rightarrow t^*(m) = m.$
2. $q < m < 1 \Rightarrow t^*(m) = 1.$
3. $m \leq q \Rightarrow t^*(m) = m.$

The first case implies a zero penalty for large positive margins. The second case reduces to the regular quadratic loss function. The third case can flip the label of a potentially mislabeled example, but still incurring a penalty of $(1-q)^2$ to keep in order with the original labeling. This way, the parameter q regulates the tolerance of negative margins.

As $q \to -\infty$, the loss becomes more and more similar to the regular convex quadratic loss function. As the other extreme, 0, is approached, the loss becomes more similar to the 0-1 loss. The optimal value of q depends on the amount of label noise in the data, and therefore it must be cross-validated.

14.6 Sparsity, Bit Depth, and Generalization Performance

Quantum boosting deviates from AdaBoost in two points. Instead of the exponential loss function in Equation 9.4, QBoost uses a quadratic loss. Moreover, it includes a regularization term to increase the sparsity of the strong learner.

$$\underset{\mathbf{w}}{\operatorname{argmin}} \left(\sum_{i=1}^{N}\left(\frac{1}{K}\mathbf{w}^\top \mathbf{h}(\mathbf{x}_i) - y_i\right)^2 + \lambda\|w\|_0\right). \tag{14.13}$$

The weights in this formulation are binary, and not real numbers as in AdaBoost. While weights could take an arbitrary value by increasing the bit depth, a few bits are, in fact, sufficient (Neven et al., 2008).

A Vapnik-Chervonenkis dimension of a strong learner is dependent on the number of weak learners included (Equation 9.3). Thus, through Vapnik's theorem (Equation 2.15), a sparser strong learner with the same training error will have a tighter upper bound on the generalization performance; hence, the regularization term in the QUBO formulation is important.

The weights are binary, which makes it easy to derive an optimal value for the λ regularization parameter, ensuring that weak classifiers are not excluded at the expense of a higher training error.

We must transform the continuous weights $w_i \in [0, 1]$ to binary variables, which is easily achieved by a binary expansion of the weights. How many bits do we need for efficient optimization? As a binary variable is associated with a qubit, and current hardware implementations have a very limited number of qubits, it is of crucial importance to use a minimal number of bits to represent the weights. The structure of the problem implies a surprisingly low number of necessary bits.

We turn to a geometric picture to gain insight into why this is the case. We map the binary expansion of a weight to a hypercube of 2^{bits} vertices, and with K weak learners, we have a total of $(2^{\text{bits}})^K$ vertices.

For a correct solution of the strong classifier, we require that

$$y_i \sum_{s=1}^{K} w_s h_s(\mathbf{x}_i) \geq 0 \qquad i = 1, \ldots, N. \tag{14.14}$$

This inequality forces us to choose weights on either side on a hyperplane in $\mathbb{R}^K$ for each training instance. The number of possible regions, K_{regions}, that can be thus generated is bound by (Orlik and Terao, 1992)

$$K_{\text{regions}} \leq \sum_{k=0}^{K} \binom{N}{k}. \tag{14.15}$$

The number of vertices in the hypercube must be at least as large as the number of possible regions. This can be relaxed by constructing the weak classifiers in a way such that only the positive quadrant is interesting, which divides the number of possible regions by about 2^N. Hence, we require

$$\frac{(2^{\text{bits}})^K}{\frac{K_{\text{regions}}}{2^K}} \geq 1. \tag{14.16}$$

Expanding this further, we get

$$\frac{(2^{\text{bits}})^K}{\frac{L_{\text{regions}}}{2^K}} = \frac{(2^{\text{bits}+1})^K}{N_{\text{regions}}} \geq \frac{(2^{\text{bits}+1})^K}{\sum_{k=0}^{K} \binom{N}{k}} \geq \frac{(2^{\text{bits}+1})^K}{\left(\frac{eN}{K}\right)^K} \geq 1. \tag{14.17}$$

We used $\sum_{k=0}^{K} \binom{N}{k} \le \left(\frac{eN}{K}\right)^K$ if $N \ge K$. From simple rearrangement, we obtain the lower bound on the number of bits:

$$\text{bits} \ge \log_2 \frac{N}{K} + \log_2 e - 1. \tag{14.18}$$

This is a remarkable result, as it shows that necessary bit precision grows only logarithmically with the ratio of training examples to weak learners. A low bit precision is essentially an implicit regularization that ensures further sparsity.

The objective function in Equation 14.4 needs slight adjustments. The continuous weights mean arbitrary scaling, but when discretizing to low bit depth such scaling is not feasible. To maintain desirable margins, we introduce a discrete scaling factor $\hat{\kappa}$. The optimal value of this parameter is unknown and it depends on the data; hence, cross-validation is necessary. Let $\dot{w}_i$ denote discretized weights. Assuming a one-bit discretization, the objective function becomes

$$\underset{\mathbf{w}}{\operatorname{argmin}} \left(\frac{\hat{\kappa}}{N} \sum_{i=1}^{K} \sum_{j=1}^{K} \dot{w}_i \dot{w}_j \left(\sum_{s=1}^{N} h_i(\mathbf{x}_s) h_j(\mathbf{x}_s) \right) + \sum_{i=1}^{K} \dot{w}_i \left(\lambda - \frac{2\hat{\kappa}}{N} \sum_{s=1}^{N} h_i(\mathbf{x}_s) y_s \right) \right). \tag{14.19}$$

This equation is indeed a valid QUBO.

Multibit weights need further auxiliary bits $\ddot{w}_j$ for each discrete variable $\dot{w}_j$ to express the regularization term. The cardinality regularization becomes

$$\lambda \|\dot{\mathbf{w}}\| = \sum_{j=1}^{K} \left\{ \phi \left(\sum_{t=1}^{d_w - 1} \dot{w}_{jt} + 1 - \dot{w}_{jd_w} \right) (1 - \ddot{w}_j) + \lambda \ddot{w}_j \right\}, \tag{14.20}$$

where $\phi > 0$ is a parameter, d_w is the bit width, and $\dot{w}_{jt}$ is the tth bit of the weight $\dot{w}_j$.

Minimizing Equation 14.20 causes the auxiliary variables $\ddot{w}_j$ to act as indicator bits when the matching variable is nonzero. This representation assumes that the most significant bit is 1 and all others are 0 if $\dot{w}_j$. If the inner sum adds up to nonzero, the parameter ϕ will add a positive value on the first term, which forces $\ddot{w}_j$ to switch to 1. This, in turn, is penalized by $\lambda \ddot{w}_j$.

Experimental results using simulated annealing and tabu search on classical hardware verified this theoretical result, showing that the error rate in 30-fold cross-validation of 1-bit precision was up to 10% better than that using double-precision weights (Neven et al., 2008). Sparsity was up to 50% better.

14.7 Mapping to Hardware

The current attempts at developing adiabatic quantum computers are not universal quantum computers. They perform quantum annealing at finite temperature on an Ising model (Boixo et al., 2013; Johnson et al., 2011).

Apart from being restricted to solving certain combinatorial optimization problems, there are additional engineering constraints to consider when implementing learning algorithms on this hardware. In manufacturing the hardware, not all connections are possible—that is, not all pairs of qubits are entangled. The connectivity is sparse, but it is known in advance. The qubits are connected in an arrangement known as a Chimera graph (McGeoch and Wang, 2013). This still put limits on the search space.

In a Chimera graph, groups of eight qubits are connected as bipartite full graphs ($K_{4,4}$). In each of these groups, the four nodes on the left side are further connected to their respective north and south neighbors in the grid. The four nodes on the right side are connected to their east and west neighbors (Figure 14.2). This way, internal nodes have a degree of six, whereas boundary nodes have a degree of five.

As part of the manufacturing process, some qubits will not be operational, or the connection between two pairs will not be functional, which further restricts graph connectivity.

To minimize the information loss, we have to find an optimal mapping between nonzero correlations in Equation 14.4 and the connections in the quantum processor. We define a graph $G = (V, E)$ to represent the actual connectivity between qubits—that is, a subgraph of the Chimera graph. We deal with the Ising model equivalent of the QUBO defined in Equation 14.5, and we map those variables to the qubit connectivity graph with a function $\phi : \{1, \ldots, n\} \mapsto V$ such that $(\phi(i), \phi(j)) \in E \Rightarrow J_{ij} \neq 0$, where n is the number of optimization variables in the QUBO.

We encode ϕ as a set of binary variables ϕ_{iq}—these indicate whether an optimization variable i is mapped to a qubit q. Naturally, we require

$$\sum_q \phi_{iq} = 1 \tag{14.21}$$

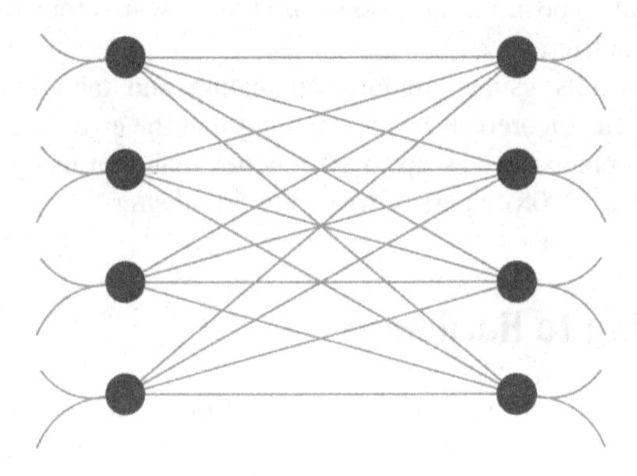

Figure 14.2 An eight-node cluster in a Chimera graph.

for all optimization variables i, and also

$$\sum_{i} \phi_{iq} \leq 1 \tag{14.22}$$

for all qubits q.

Minimizing the information loss, we seek to maximize the magnitude of J_{ij} mapped to qubit edges—that is, we are seeking

$$\underset{\phi}{\operatorname{argmax}} \sum_{i>j} \sum_{(q,q')\in E} |J_{ij}|\phi_{iq}\phi_{iq'}, \tag{14.23}$$

with the constraints applying to ϕ in Equations 14.21 and 14.22.

This problem itself is in fact NP-hard, being a variant of the quadratic assignment problem. It must be solved at each invocation of the quantum hardware; hence, a fast heuristic is necessary to approximate the optimum. The following algorithm finds an approximation in $O(n)$ time complexity (Neven et al., 2009).

Initially, let $i_1 = \operatorname{argmax}_i \sum_{j<i} |J_{ji}| + \sum_{j>i} |J_{ij}|$—that is i_1 is the row or column index of J with the highest sum of magnitudes. We assign i_1 to one of the qubit vertices of the highest degree.

For the generic step, we already have a set $\{i_1, \ldots, i_k\}$ such that $\phi(i_j) = q_j$. To assign the next $i_{k+1} \notin \{i_1, \ldots, i_k\}$ to an unmapped qubit q_{k+1}, we need to maximize the sum of all $|J_{i_{k+1}i_j}|$ and $|J_{i_j i_{k+1}}|$ over all $j \in 1, \ldots, k$, where $\{q_j, q_{k+1}\} \in E$.

This greedy heuristic reportedly performs well, mapping about 11% of the total absolute edge weight $\sum_{i,j} |J_{ij}|$ of a fully connected random Ising model into actual hardware connectivity in a few milliseconds, whereas a tabu heuristic on the same problem performs only marginally better, with a run time in the range of a few minutes (Neven et al., 2009).

Sparse qubit connectivity is not the only problem with current quantum hardware implementations. While the optimum is achieved in the ground state at absolute zero, these systems run at nonzero temperature, at around 20-40 mK. This is significant at the scales of an Ising model, and thermally excited states are observed in experiments. This also introduces problems on the minimum gap. Solving this issue requires multiple runs on the same problem, and finally choosing the result with the lowest energy. For a 128-qubit configuration, obtaining m solutions to the same problem takes approximately $900 + 100m$ milliseconds, with $m = 32$ giving good performance (Neven et al., 2009).

A further problem is that the number of candidate weak classifiers may exceed the number of variables that can be handled in a single optimization run on the hardware. We refer to such situations as large-scale training (Neven et al., 2012). It is also possible that the final selected weak classifiers exceed the number of available variables.

An iterative and piecewise approach deals with these cases in which at each iteration a subset of weak classifiers is selected via global optimization. Let Q denote the number of weak classifiers the hardware can accommodate at a time, let T_{outer} denote the total number of selected weak learners, and let $c(\mathbf{x})$ denote the current weighted sum of weak learners. Algorithm 8 describes the extension of QBoost that can handle problems of arbitrary size.

ALGORITHM 8 QBoost outer loop

Require: Training and validation data, dictionary of weak classifiers
Ensure: Strong classifier
  Initialize weight distribution d_{outer} over training samples as uniform distribution $\forall s : d_{\text{outer}}(s) = 1/K$
  Set $T_{\text{outer}} \leftarrow 0$ and $c(\mathbf{x}) \leftarrow 0$
  repeat
    Run Algorithm 7 with d initialized from current d_{outer} and using an objective function that takes into account the current $c(\mathbf{x})$: $\mathbf{w}^\star = \text{argmin}_{\mathbf{w}}(\sum_{s=1}^{K}[(c(\mathbf{x}_s) + \sum_{i=1}^{Q} w_i h_i(\mathbf{x}_s))/(T_{\text{outer}} + Q) - y_s]^2 + \lambda\|\mathbf{w}\|_0)$.
    Set $T_{\text{outer}} \leftarrow T_{\text{outer}} + \|\mathbf{w}^\star\|_0$ and $c(\mathbf{x}) \leftarrow c(\mathbf{x}) + \sum_{i=1}^{Q} w_i^\star h_i(\mathbf{x})$
    Construct a strong classifier $H(\mathbf{x}) = \text{sign}(c(\mathbf{x}))$
    Update weights $d_{\text{outer}}(s) = d_{\text{outer}}(s)(\sum_{t=1}^{T_{\text{outer}}} h_t(x)/T_{\text{outer}} - y_s)^2$
    Normalize $d_{\text{outer}}(s) = d_{\text{outer}}(s)/\sum_{s=1}^{S} d_{\text{outer}}(s)$
  until validation error E_{val} stops decreasing

QBoost thus considers a group of Q weak classifiers at a time—Q is the limit imposed by the constraints—and finds a subset with the lowest empirical risk on Q. If the error reaches the optimum on Q, this means that more weak classifiers are necessary to decrease the error rate further. At this point, the algorithm changes the working set Q, leaving earlier selected weak classifiers invariant.

Compared with the best known implementations on classical results, McGeoch and Wang (2013) found that the actual computational time was shorter on adiabatic quantum hardware for a QUBO, but it finished calculations in approximately the same time in other optimization problems. This was a limited experimental validation using specific data sets. Further research into computational time showed that the optimal time for annealing was underestimated, and there was no evidence of quantum speedup on an Ising model (Rønnow et al., 2014).

Another problem with the current implementation of adiabatic quantum computers is that demonstrating quantum effects is inconclusive. There is evidence for correlation between quantum annealing in an adiabatic quantum processor and simulated quantum annealing (Boixo et al., 2014), and there are signs of entanglement during annealing (Lanting et al., 2014). Yet, classical models for this quantum processor are still not ruled out (Shin et al., 2014).

14.8 Computational Complexity

Time complexity derives from how long the adiabatic process must take to find the global optimum with high probability. The quantum adiabatic theorem states that the adiabatic evolution of the system depends on the time $\tau = t_1 - t_0$ during which the change takes place. This time is proportional to a power law:

$$\tau \propto g_{\min}^{-\delta}, \tag{14.24}$$

where $g_{\min}$ is the minimum gap in the lowest-energy eigenstates of the system Hamiltonian, and δ depends on the parameter λ and the distribution of eigenvalues at higher energy levels. For instance, δ may equal 1 (Schaller et al., 2006), 2 (Farhi et al., 2000), or, in certain circumstances, even 3 (Lidar et al., 2009). To understand the efficiency of adiabatic quantum computing, we need to analyze $g_{\min}$, but in practice, this is a difficult task (Amin and Choi, 2009).

A few cases have analytic solutions, but in general, we have to resort to numerical methods such as exact diagonalization and quantum Monte Carlo methods. These are limited to small problem sizes and they offer little insight into why the gap is of a particular size (Young et al., 2010).

For the Ising model, the gap size scales linearly with the number of variables in the problem (Neven et al., 2012). Together with Equation 14.24, this implies a polynomial time complexity for finding the optimum of a QUBO. Yet, in other cases, the Hamiltonian is sensitive to perturbations, leading to exponential changes in the gap as the problem size increases (Amin and Choi, 2009). In some cases, we overcome such problems by randomly modifying the base Hamiltonian, and running the computation several times, always leading to the target Hamiltonian. For instance, we can modify the base Hamiltonian in Equation 14.8 by adding n random variables c_i:

$$H_{\mathrm{B}} = \sum_{i=1}^{n} \frac{c_i(1 - \sigma_i^x)}{2}. \tag{14.25}$$

Since some Hamiltonians are sensitive to the initial conditions, this random perturbation may reduce the small gap that causes long run times (Farhi et al., 2011).

Even if finding the global optimum takes exponential time, early exit might yield good results. Owing to quantum tunneling, the approximate solutions can still be better than those obtained by classical algorithms (Neven et al., 2012).

It is an open question how the gapless formulation of the adiabatic theorem influences time complexity.

Bibliography

Abu-Mostafa, Y., St. Jacques, J.-M., 1985. Information capacity of the Hopfield model. IEEE Trans. Inf. Theory 31(4), 461–464.

Acín, A., Jané, E., Vidal, G., 2001. Optimal estimation of quantum dynamics. Phys. Rev. A 64, 050302.

Aerts, D., Czachor, M., 2004. Quantum aspects of semantic analysis and symbolic artificial intelligence. J. Phys. A Math. Gen. 37, L123-L132.

Aharonov, D., Van Dam, W., Kempe, J., Landau, Z., Lloyd, S., Regev, O., 2004. Adiabatic quantum computation is equivalent to standard quantum computation. In: Proceedings of FOCS-04, 45th Annual IEEE Symposium on Foundations of Computer Science.

Aïmeur, E., Brassard, G., Gambs, S., 2013. Quantum speed-up for unsupervised learning. Mach. Learn. 90(2), 261–287.

Altaisky, M.V., 2001. Quantum neural network. arXiv:quant-ph/0107012.

Altepeter, J.B., Branning, D., Jeffrey, E., Wei, T., Kwiat, P.G., Thew, R.T., O'Brien, J.L., Nielsen, M.A., White, A.G., 2003. Ancilla-assisted quantum process tomography. Phys. Rev. Lett. 90(19), 193601.

Amin, M.H.S., Choi, V., 2009. First-order quantum phase transition in adiabatic quantum computation. Phys. Rev. A 80, 062326.

Amin, M.H.S., Truncik, C.J.S., Averin, D.V., 2009. Role of single-qubit decoherence time in adiabatic quantum computation. Phys. Rev. A 80, 022303.

Angluin, D., 1988. Queries and concept learning. Mach. Learn. 2(4), 319–342.

Anguita, D., Ridella, S., Rivieccio, F., Zunino, R., 2003. Quantum optimization for training support vector machines. Neural Netw. 16(5), 763–770.

Ankerst, M., Breunig, M., Kriegel, H., Sander, J., 1999. OPTICS: ordering points to identify the clustering structure. In: Proceedings of SIGMOD-99, International Conference on Management of Data, pp. 49–60.

Asanovic, K., Bodik, R., Catanzaro, B., Gebis, J., Husbands, P., Keutzer, K., Patterson, D., Plishker, W., Shalf, J., Williams, S., 2006. The landscape of parallel computing research: a view from Berkeley. Technical Report, University of California at Berkeley.

Aspect, A., Dalibard, J., Roger, G., 1982. Experimental test of Bell's inequalities using time-varying analyzers. Phys. Rev. Lett. 49, 1804–1807.

Atici, A., Servedio, R.A., 2005. Improved bounds on quantum learning algorithms. Quantum Inf. Process. 4(5), 355–386.

Avron, J.E., Elgart, A., 1999. Adiabatic theorem without a gap condition. Commun. Math. Phys. 203(2), 445–463.

Bacon, D., van Dam, W., 2010. Recent progress in quantum algorithms. Commun. ACM 53(2), 84–93.

Beckmann, N., Kriegel, H., Schneider, R., Seeger, B., 1990. The R*-tree: an efficient and robust access method for points and rectangles. SIGMOD Rec. 19(2), 322–331.

Behrman, E.C., Niemel, J., Steck, J.E., Skinner, S.R., 1996. A quantum dot neural network. In: Proceedings of PhysComp-96, 4th Workshop on Physics of Computation, pp. 22–28.

Behrman, E.C., Nash, L., Steck, J.E., Chandrashekar, V., Skinner, S.R., 2000. Simulations of quantum neural networks. Inform. Sci. 128(3), 257–269.

Bell, J., 1964. On the Einstein Podolsky Rosen paradox. Physics 195-200(3), 1.

Bengio, Y., LeCun, Y., 2007. Scaling learning algorithms towards AI. In: Bottou, L., Chapelle, O., DeCoste, D., Weston, J. (Eds.), Large-Scale Kernel Machines. MIT Press, Cambridge, MA, pp. 321–360.

Bennett, C., Bernstein, E., Brassard, G., Vazirani, U., 1997. Strengths and weaknesses of quantum computing. SIAM J. Comput. 26(5), 1510–1523.

Berchtold, S., Keim, D.A., Kriegel, H.-P., 1996. The X-tree: an index structure for high-dimensional data. In: Vijayaraman, T.M., Buchmann, A.P., Mohan, C., Sarda, N.L. (Eds.), Proceedings of VLDB-96, 22th International Conference on Very Large Data Bases. Morgan Kaufmann Publishers, San Francisco, CA, pp. 28–39.

Berry, D.W., Ahokas, G., Cleve, R., Sanders, B.C., 2007. Efficient quantum algorithms for simulating sparse Hamiltonians. Commun. Math. Phys. 270(2), 359–371.

Bisio, A., Chiribella, G., D'Ariano, G.M., Facchini, S., Perinotti, P., 2010. Optimal quantum learning of a unitary transformation. Phys. Rev. A 81(3), 032324.

Bisio, A., D'Ariano, G.M., Perinotti, P., Sedlák, M., 2011. Quantum learning algorithms for quantum measurements. Phys. Lett. A 375, 3425–3434.

Blekas, K., Lagaris, I., 2007. Newtonian clustering: an approach based on molecular dynamics and global optimization. Pattern Recognit. 40(6), 1734–1744.

Blumer, A., Ehrenfeucht, A., Haussler, D., Warmuth, M.K., 1989. Learnability and the Vapnik-Chervonenkis dimension. J. ACM 36(4), 929–965.

Boixo, S., Albash, T., Spedalieri, F., Chancellor, N., Lidar, D., 2013. Experimental signature of programmable quantum annealing. Nat. Commun. 4, 2067.

Boixo, S., Rønnow, T.F., Isakov, S.V., Wang, Z., Wecker, D., Lidar, D.A., Martinis, J.M., Troyer, M., 2014. Evidence for quantum annealing with more than one hundred qubits. Nat. Phys. 10(3), 218–224.

Bonner, R., Freivalds, R., 2002. A survey of quantum learning. In: Bonner, R., Freivalds, R. (Eds.), Proceedings of QCL-02, 3rd International Workshop on Quantum Computation and Learning. Mälardalen University Press, Västerås and Eskilstuna.

Born, M., Fock, V., 1928. Beweis des adiabatensatzes. Z. Phys. 51(3-4), 165–180.

Bradley, P.S., Fayyad, U.M., 1998. Refining initial points for K-means clustering. In: Proceedings of ICML-98, 15th International Conference on Machine Learning. Morgan Kaufmann, San Francisco, CA, pp. 91–99.

Brassard, G., Cleve, R., Tapp, A., 1999. Cost of exactly simulating quantum entanglement with classical communication. Phys. Rev. Lett. 83, 1874–1877.

Breiman, L., 1996. Bagging predictors. Mach. Learn. 24(2), 123–140.

Breiman, L., 2001. Random forests. Mach. Learn. 45(1), 5–32.

Bruza, P., Cole, R., 2005. Quantum logic of semantic space: an exploratory investigation of context effects in practical reasoning. In: Artemov, S., Barringer, H., d'Avila Garcez, A.S., Lamb, L., Woods, J. (Eds.), We Will Show Them: Essays in Honour of Dov Gabbay. College Publications, London, UK, pp. 339–361.

Bshouty, N.H., Jackson, J.C., 1995. Learning DNF over the uniform distribution using a quantum example oracle. In: Proceedings of COLT-95, 8th Annual Conference on Computational Learning Theory, pp. 118–127.

Buhrman, H., Cleve, R., Watrous, J., De Wolf, R., 2001. Quantum fingerprinting. Phys. Rev. Lett. 87(16), 167902.

Burges, C., 1998. A tutorial on support vector machines for pattern recognition. Data Min. Knowl. Discov. 2(2), 121–167.

Chatterjee, A., Bhowmick, S., Raghavan, P., 2008. FAST: force-directed approximate subspace transformation to improve unsupervised document classification. In: Proceedings of 6th Text Mining Workshop Held in Conjunction with SIAM International Conference on Data Mining.

Childs, A.M., Farhi, E., Preskill, J., 2001. Robustness of adiabatic quantum computation. Phys. Rev. A 65, 012322.

Chiribella, G., D'Ariano, G.M., Sacchi, M.F., 2005. Optimal estimation of group transformations using entanglement. Phys. Rev. A 72(4), 042338.

Chiribella, G., 2011. Group theoretic structures in the estimation of an unknown unitary transformation. J. Phys. Conf. Ser. 284(1), 012001.

Choi, M.-D., 1975. Completely positive linear maps on complex matrices. Linear Algebra Appl. 10(3), 285–290.

Chuang, I.L., Nielsen, M.A., 1997. Prescription for experimental determination of the dynamics of a quantum black box. J. Mod. Opt. 44(11-12), 2455–2467.

Ciaccia, P., Patella, M., Zezula, P., 1997. M-tree: an efficient access method for similarity search in metric spaces. In: Proceedings of VLDB-97, 23rd International Conference on Very Large Data Bases, pp. 426–435.

Clauser, J.F., Horne, M.A., Shimony, A., Holt, R.A., 1969. Proposed experiment to test local hidden-variable theories. Phys. Rev. Lett. 23, 880–884.

Cohen, W., Singer, Y., 1996. Context-sensitive learning methods for text categorization. In: Proceedings of SIGIR-96, 19th International Conference on Research and Development in Information Retrieval, pp. 307–315.

Cohen-Tannoudji, C., Diu, B., Laloë, F., 1996. Quantum Mechanics. John Wiley & Sons, New York.

Collobert, R., Sinz, F., Weston, J., Bottou, L., 2006. Trading convexity for scalability. In: Proceedings of ICML-06, 23rd International Conference on Machine Learning, pp. 201–208.

Copas, J.B., 1983. Regression, prediction and shrinkage. J. R. Stat. Soc. Ser. B Methodol. 45, 311–354.

Cox, T., Cox, M., 1994. Multidimensional Scaling. Chapman and Hall, Boca Raton.

Cox, D.R., 2006. Principles of Statistical Inference. Cambridge University Press, Cambridge.

Cristianini, N., Shawe-Taylor, J., 2000. An Introduction to Support Vector Machines and Other Kernel-Based Learning Methods. Cambridge University Press, Cambridge.

Cui, X., Gao, J., Potok, T., 2006. A flocking based algorithm for document clustering analysis. J. Syst. Archit. 52(8), 505–515.

D'Ariano, G.M., Lo Presti, P., 2003. Imprinting complete information about a quantum channel on its output state. Phys. Rev. Lett. 91, 047902.

De Silva, V., Tenenbaum, J., 2003. Global versus local methods in nonlinear dimensionality reduction. Adv. Neural Inf. Process. Syst. 15, 721–728.

Deerwester, S., Dumais, S., Furnas, G., Landauer, T., Harshman, R., 1990. Indexing by latent semantic analysis. J. Am. Soc. Inf. Sci. 41(6), 391–407.

Demiriz, A., Bennett, K.P., Shawe-Taylor, J., 2002. Linear programming boosting via column generation. Mach. Learn. 46(1-3), 225–254.

Denchev, V.S., Ding, N., Vishwanathan, S., Neven, H., 2012. Robust classification with adiabatic quantum optimization. In: Proceedings of ICML-2012, 29th International Conference on Machine Learning.

Deutsch, D., 1985. Quantum theory, the Church-Turing principle and the universal quantum computer. Proc. R. Soc. A 400(1818), 97–117.

Dietterich, T.G., 2000. An experimental comparison of three methods for constructing ensembles of decision trees: bagging, boosting, and randomization. Mach. Learn. 40(2), 139–157.

Ding, C., He, X., 2004. K-means clustering via principal component analysis. In: Proceedings of ICML-04, 21st International Conference on Machine Learning, pp. 29–37.

Dong, D., Chen, C., Li, H., Tarn, T.-J., 2008. Quantum reinforcement learning. IEEE Trans. Syst. Man Cybern. B Cybern. 38(5), 1207–1220.

Drucker, H., Burges, C.J., Kaufman, L., Smola, A., Vapnik, V., 1997. Support vector regression machines. Adv. Neural Inf. Process. Syst. 10, 155–161.

Duan, L.-M., Guo, G.-C., 1998. Probabilistic cloning and identification of linearly independent quantum states. Phys. Rev. Lett. 80, 4999–5002.

Duffy, N., Helmbold, D., 2000. Potential boosters? Adv. Neural Inf. Process. Syst. 13, 258–264.

Durr, C., Hoyer, P., 1996. A quantum algorithm for finding the minimum. arXiv:quant-ph/9607014.

Efron, B., 1979. Bootstrap methods: another look at the jackknife. Ann. Stat. 7(1), 1–26.

El-Yaniv, R., Pechyony, D., 2007. Transductive Rademacher complexity and its applications. In: Bshouty, N.H., Gentile, C. (Eds.), Proceedings of COLT-07, 20th Annual Conference on Learning Theory. Springer, Berlin, pp. 157–171.

Erhan, D., Bengio, Y., Courville, A., Manzagol, P.-A., Vincent, P., Bengio, S., 2010. Why does unsupervised pre-training help deep learning? J. Mach. Learn. Res. 11, 625–660.

Ertekin, S., Bottou, L., Giles, C.L., 2011. Nonconvex online support vector machines. IEEE Trans. Pattern Anal. Mach. Intell. 33(2), 368–381.

Ester, M., Kriegel, H., Sander, J., Xu, X., 1996. A density-based algorithm for discovering clusters in large spatial databases with noise. In: Proceedings of SIGKDD-96, 2nd International Conference on Knowledge Discovery and Data Mining, vol. 96, pp. 226–231.

Ezhov, A.A., Ventura, D., 2000. Quantum neural networks. In: Kasabov, N. (Ed.), Future Directions for Intelligent Systems and Information Sciences, Studies in Fuzziness and Soft Computing. Physica-Verlag HD, Heidelberg, pp. 213–235.

Farhi, E., Goldstone, J., Gutmann, S., Sipser, M., 2000. Quantum computation by adiabatic evolution. arXiv:quant-ph/0001106.

Farhi, E., Goldston, J., Gosset, D., Gutmann, S., Meyer, H.B., Shor, P., 2011. Quantum adiabatic algorithms, small gaps, and different paths. Quantum Inf. Comput. 11(3), 181–214.

Fayngold, M., Fayngold, V., 2013. Quantum Mechanics and Quantum Information. Wiley-VCH, Weinheim.

Feldman, V., Guruswami, V., Raghavendra, P., Wu, Y., 2012. Agnostic learning of monomials by halfspaces is hard. SIAM J. Comput. 41(6), 1558–1590.

Feynman, R.P., 1982. Simulating physics with computers. Int. J. Theor. Phys. 21(6), 467–488.

Finnila, A., Gomez, M., Sebenik, C., Stenson, C., Doll, J., 1994. Quantum annealing: a new method for minimizing multidimensional functions. Chem. Phys. Lett. 219(5-6), 343–348.

Freund, Y., Schapire, R.E., 1997. A decision-theoretic generalization of on-line learning and an application to boosting. J. Comput. Syst. Sci. 55(1), 119–139.

Friedman, J., Hastie, T., Tibshirani, R., 2000. Additive logistic regression: a statistical view of boosting. Ann. Stat. 28(2), 337–407.

Friedman, J.H., 2001. Greedy function approximation: gradient boosting machine. Ann. Stat. 29(5), 1189–1232.

Fuchs, C., 2002. Quantum mechanics as quantum information (and only a little more). arXiv:quant-ph/0205039.

Gambs, S., 2008. Quantum classification. arXiv:0809.0444.

Gammerman, A., Vovk, V., Vapnik, V., 1998. Learning by transduction. In: Proceedings of UAI-98, 14th Conference on Uncertainty in Artificial Intelligence, pp. 148–155.
Gardner, E., 1988. The space of interactions in neural network models. J. Phys. A Math. Gen. 21(1), 257.
Gavinsky, D., 2012. Quantum predictive learning and communication complexity with single input. Quantum Inf. Comput. 12(7-8), 575–588.
Giovannetti, V., Lloyd, S., Maccone, L., 2008. Quantum random access memory. Phys. Rev. Lett. 100(16), 160501.
Glover, F., 1989. Tabu search—part I. ORSA J. Comput. 1(3), 190–206.
Goldberg, D.E., 1989. Genetic Algorithms in Search, Optimization, and Machine Learning. Addison-Wesley Professional, Upper Saddle River, NJ.
Grover, L.K., 1996. A fast quantum mechanical algorithm for database search. In: Proceedings of STOC0-96, 28th Annual ACM Symposium on Theory of Computing, pp. 212–219.
Guţă, M., Kotłowski, W., 2010. Quantum learning: asymptotically optimal classification of qubit states. New J. Phys. 12(12), 123032.
Gupta, S., Zia, R., 2001. Quantum neural networks. J. Comput. Syst. Sci. 63(3), 355–383.
Guyon, I., Elisseefi, A., Kaelbling, L., 2003. An introduction to variable and feature selection. J. Mach. Learn. Res. 3(7-8), 1157–1182.
Han, J., Kamber, M., Pei, J., 2012. Data Mining: Concepts and Techniques, third ed. Morgan Kaufmann, Burlington, MA.
Härdle, W.K., 1990. Applied Nonparametric Regression. Cambridge University Press, Cambridge.
Harrow, A.W., Hassidim, A., Lloyd, S., 2009. Quantum algorithm for linear systems of equations. Phys. Rev. Lett. 103(15), 150502.
Hastie, T., Tibshirani, R., Friedman, J., 2008. The Elements of Statistical Learning: Data Mining, Inference, and Prediction, second ed. Springer.
Haussler, D., 1992. Decision theoretic generalizations of the PAC model for neural net and other learning applications. Inf. Comput. 100(1), 78–150.
Hinton, G., Deng, L., Yu, D., Dahl, G.E., Mohamed, A.R., Jaitly, N., Senior, A., Vanhoucke, V., Nguyen, P., Sainath, T.N., et al., 2012. Deep neural networks for acoustic modeling in speech recognition: the shared views of four research groups. IEEE Signal Process. Mag. 29(6), 82–97.
Holevo, A., 1982. Probabilistic and Statistical Aspects of Quantum Theory. North-Holland Publishing Company, Amsterdam.
Holte, R., 1993. Very simple classification rules perform well on most commonly used datasets. Mach. Learn. 11(1), 63–90.
Hopfield, J.J., 1982. Neural networks and physical systems with emergent collective computational abilities. Proc. Natl. Acad. Sci. U.S.A. 79(8), 2554–2558.
Hornik, K., Stinchcombe, M., White, H., 1989. Multilayer feedforward networks are universal approximators. Neural Netw. 2(5), 359–366.
Horodecki, M., Horodecki, P., Horodecki, R., 1996. Separability of mixed states: necessary and sufficient conditions. Phys. Lett. A 223(1), 1–8.
Hsu, C., Lin, C., 2002. A comparison of methods for multiclass support vector machines. IEEE Trans. Neural Netw. 13(2), 415–425.
Huang, G.-B., Babri, H.A., 1998. Upper bounds on the number of hidden neurons in feedforward networks with arbitrary bounded nonlinear activation functions. IEEE Trans. Neural Netw. 9(1), 224–229.
Huang, G.-B., 2003. Learning capability and storage capacity of two-hidden-layer feedforward networks. IEEE Trans. Neural Netw. 14(2), 274–281.

Huang, G.-B., Zhu, Q.-Y., Siew, C.-K., 2006. Extreme learning machine: theory and applications. Neurocomputing 70(1-3), 489–501.

Iba, W., Langley, P., 1992. Induction of one-level decision trees. In: Proceedings of ML-92, 9th International Workshop on Machine Learning, pp. 233–240.

Ito, M., Miyoshi, T., Masuyama, H., 2000. The characteristics of the torus self organizing map. Faji Shisutemu Shinpojiumu Koen Ronbunshu 16, 373–374.

Jamiołkowski, A., 1972. Linear transformations which preserve trace and positive semidefiniteness of operators. Rep. Math. Phys. 3(4), 275–278.

Joachims, T., 1998. Text categorization with support vector machines: learning with many relevant features. In: Proceedings of ECML-98, 10th European Conference on Machine Learning, pp. 137–142.

Joachims, T., 2006. Training linear SVMs in linear time. In: Proceedings of SIGKDD-06, 12th International Conference on Knowledge Discovery and Data Mining, pp. 217–226.

Johnson, M., Amin, M., Gildert, S., Lanting, T., Hamze, F., Dickson, N., Harris, R., Berkley, A., Johansson, J., Bunyk, P., et al., 2011. Quantum annealing with manufactured spins. Nature 473(7346), 194–198.

Jolliffe, I., 1989. Principal Component Analysis. Springer, New York, NY.

Katayama, K., Narihisa, H., 2001. Performance of simulated annealing-based heuristic for the unconstrained binary quadratic programming problem. Eur. J. Oper. Res. 134(1), 103–119.

Kendon, V.M., Nemoto, K., Munro, W.J., 2010. Quantum analogue computing. Philos. Trans. R. Soc. A Math. Phys. Eng. Sci. 368(1924), 3609–3620.

Kennedy, J., Eberhart, R., 1995. Particle swarm optimization. In: Proceedings of ICNN-95, International Conference on Neural Networks, pp. 1942–1948.

Khrennikov, A., 2010. Ubiquitous Quantum Structure: From Psychology to Finance. Springer-Verlag, Heidelberg.

Kitto, K., 2008. Why quantum theory? In: Proceedings of QI-08, 2nd International Symposium on Quantum Interaction, pp. 11–18.

Kohavi, R., John, G., 1997. Wrappers for feature subset selection. Artif. Intell. 97(1-2), 273–324.

Kondor, R., Lafferty, J., 2002. Diffusion kernels on graphs and other discrete input spaces. In: Proceedings of ICML-02, 19th International Conference on Machine Learning, pp. 315–322.

Kraus, B., 2013. Topics in quantum information. In: DiVincenzo, D. (Ed.), Lecture Notes of the 44th IFF Spring School "Quantum Information Processing". Forschungszentrum Jülich.

Kriegel, H.-P., Kröger, P., Sander, J., Zimek, A., 2011. Density-based clustering. Wiley Interdiscip. Rev. Data Min. Knowl. Discov. 1(3), 231–240.

Kruskal, W., 1988. Miracles and statistics: the casual assumption of independence. J. Am. Stat. Assoc. 83(404), 929–940.

Kuncheva, L.I., Whitaker, C.J., 2003. Measures of diversity in classifier ensembles and their relationship with the ensemble accuracy. Mach. Learn. 51(2), 181–207.

Laarhoven, P.J., Aarts, E.H., 1987. Simulated Annealing: Theory and Applications. Reidel Publishing Company, The Netherlands.

Lan, M., Tan, C.L., Su, J., Lu, Y., 2009. Supervised and traditional term weighting methods for automatic text categorization. IEEE Trans. Pattern Anal. Mach. Intell. 31(4), 721–735.

Langley, P., Sage, S., 1994. Induction of selective Bayesian classifiers. In: de Mantaras, R., Poole, D. (Eds.), Proceedings of UAI-94, 10th Conference on Uncertainty in Artificial Intelligence, pp. 399–406.

Langley, P., Sage, S., 1994. Oblivious decision trees and abstract cases. In: Working Notes of the AAAI-94 Workshop on Case-Based Reasoning, pp. 113–117.

Lanting, T., Przybysz, A.J., Smirnov, A.Y., Spedalieri, F.M., Amin, M.H., Berkley, A.J., Harris, R., Altomare, F., Boixo, S., Bunyk, P., Dickson, N., Enderud, C., Hilton, J.P., Hoskinson, E., Johnson, M.W., Ladizinsky, E., Ladizinsky, N., Neufeld, R., Oh, T., Perminov, I., Rich, C., Thom, M.C., Tolkacheva, E., Uchaikin, S., Wilson, A.B., Rose, G., 2014. Entanglement in a quantum annealing processor. arXiv:1401.3500.

Larkey, L., Croft, W., 1996. Combining classifiers in text categorization. In: Proceedings of SIGIR-96, 19th International Conference on Research and Development in Information Retrieval, pp. 289–297.

Law, M., Zhang, N., Jain, A., 2004. Nonlinear manifold learning for data stream. In: Proceedings of ICDM-04, 4th IEEE International Conference on Data Mining, pp. 33–44.

Law, M., Jain, A., 2006. Incremental nonlinear dimensionality reduction by manifold learning. IEEE Trans. Pattern Anal. Mach. Intell. 28(3), 377–391.

Leung, D.W., 2003. Choi's proof as a recipe for quantum process tomography. J. Math. Phys. 44, 528.

Lewenstein, M., 1994. Quantum perceptrons. J. Mod. Opt. 41(12), 2491–2501.

Lewis, D., Ringuette, M., 1994. A comparison of two learning algorithms for text categorization. In: Proceedings of SDAIR-94, 3rd Annual Symposium on Document Analysis and Information Retrieval, pp. 81–93.

Lidar, D.A., Rezakhani, A.T., Hamma, A., 2009. Adiabatic approximation with exponential accuracy for many-body systems and quantum computation. J. Math. Phys. 50, 102106.

Lin, H., Lin, C., 2003. A study on sigmoid kernels for SVM and the training of non-PSD kernels by SMO-type methods. Technical Report, Department of Computer Science, National Taiwan University.

Lin, T., Zha, H., 2008. Riemannian manifold learning. IEEE Trans. Pattern Anal. Mach. Intell. 30(5), 796.

Lloyd, S., 1996. Universal quantum simulators. Science 273(5278), 1073–1078.

Lloyd, S., Mohseni, M., Rebentrost, P., 2013. Quantum algorithms for supervised and unsupervised machine learning. arXiv:1307.0411.

Lloyd, S., Mohseni, M., Rebentrost, P., 2013. Quantum principal component analysis. arXiv:1307.0401.

Lodhi, H., Saunders, C., Shawe-Taylor, J., Cristianini, N., Watkins, C., Scholkopf, B., 2002. Text classification using string kernels. J. Mach. Learn. Res. 2(3), 419–444.

Long, P.M., Servedio, R.A., 2010. Random classification noise defeats all convex potential boosters. Mach. Learn. 78(3), 287–304.

Loo, C.K., Peruš, M., Bischof, H., 2004. Associative memory based image and object recognition by quantum holography. Open Syst. Inf. Dyn. 11(03), 277–289.

Lu, H., Setiono, R., Liu, H., 1996. Effective data mining using neural networks. IEEE Trans. Knowl. Data Eng. 8(6), 957–961.

MacKay, D.J.C., 2005. Information Theory, Inference & Learning Algorithms, fourth ed. Cambridge University Press, Cambridge.

Manju, A., Nigam, M., 2012. Applications of quantum inspired computational intelligence: a survey. Artif. Intell. Rev. 42(1), 79–156.

Manwani, N., Sastry, P., 2013. Noise tolerance under risk minimization. IEEE Trans. Cybern. 43(3), 1146–1151.

Masnadi-Shirazi, H., Vasconcelos, N., 2008. On the design of loss functions for classification: theory, robustness to outliers, and SavageBoost. Adv. Neural Inf. Process. Syst. 21, 1049–1056.

Masnadi-Shirazi, H., Mahadevan, V., Vasconcelos, N., 2010. On the design of robust classifiers for computer vision. In: Proceedings of CVPR-10, IEEE Conference on Computer Vision and Pattern Recognition, pp. 779–786.

Mason, L., Baxter, J., Bartlett, P., Frean, M., 1999. Boosting algorithms as gradient descent in function space. Adv. Neural Inf. Process. Syst. 11, 512–518.

McGeoch, C.C., Wang, C., 2013. Experimental evaluation of an adiabatic quantum system for combinatorial optimization. In: Proceedings of CF-13, ACM International Conference on Computing Frontiers, pp. 23:1-23:11.

Minsky, M., Papert, S., 1969. Perceptrons: An Introduction to Computational Geometry. MIT Press, Cambridge, MA.

Mirsky, L., 1960. Symmetric gage functions and unitarily invariant norms. Q. J. Math. 11, 50–59.

Mishra, N., Oblinger, D., Pitt, L., 2001. Sublinear time approximate clustering. In: Proceedings of SODA-01, 12th Annual ACM-SIAM Symposium on Discrete Algorithms, pp. 439–447.

Mitchell, T., 1997. Machine Learning. McGraw-Hill, New York, NY.

Mohseni, M., Rezakhani, A.T., Lidar, D.A., 2008. Quantum-process tomography: resource analysis of different strategies. Phys. Rev. A 77, 032322.

Narayanan, A., Menneer, T., 2000. Quantum artificial neural network architectures and components. Inform. Sci. 128(3-4), 231–255.

Neigovzen, R., Neves, J.L., Sollacher, R., Glaser, S.J., 2009. Quantum pattern recognition with liquid-state nuclear magnetic resonance. Phys. Rev. A 79, 042321.

Neven, H., Denchev, V.S., Rose, G., Macready, W.G., 2008. Training a binary classifier with the quantum adiabatic algorithm. arXiv:0811.0416.

Neven, H., Denchev, V.S., Drew-Brook, M., Zhang, J., Macready, W.G., Rose, G., 2009. Binary classification using hardware implementation of quantum annealing. In: Demonstrations at NIPS-09, 24th Annual Conference on Neural Information Processing Systems, pp. 1–17.

Neven, H., Denchev, V.S., Rose, G., Macready, W.G., 2012. Qboost: large scale classifier training with adiabatic quantum optimization. In: Proceedings of ACML-12, 4th Asian Conference on Machine Learning, pp. 333–348.

Onclinx, V., Wertz, V., Verleysen, M., 2009. Nonlinear data projection on non-Euclidean manifolds with controlled trade-off between trustworthiness and continuity. Neurocomputing 72(7-9), 1444–1454.

Oppenheim, J., Wehner, S., 2010. The uncertainty principle determines the nonlocality of quantum mechanics. Science 330(6007), 1072–1074.

Orlik, P., Terao, H., 1992. Arrangements of Hyperplanes. Springer, Heidelberg.

Orponen, P., 1994. Computational complexity of neural networks: a survey. Nordic J. Comput. 1(1), 94–110.

Palubeckis, G., 2004. Multistart tabu search strategies for the unconstrained binary quadratic optimization problem. Ann. Oper. Res. 131(1-4), 259–282.

Park, H.-S., Jun, C.-H., 2009. A simple and fast algorithm for K-medoids clustering. Expert Syst. Appl. 36(2), 3336–3341.

Platt, J., 1999. Fast training of support vector machines using sequential minimal optimization. In: Schölkopf, B., Burges, C., Smola, A. (Eds.), Advances in Kernel Methods: Support Vector Learning. MIT Press, pp. 185–208.

Polikar, R., 2006. Ensemble based systems in decision making. IEEE Circuits Syst. Mag. 6(3), 21–45.

Pothos, E.M., Busemeyer, J.R., 2013. Can quantum probability provide a new direction for cognitive modeling? Behav. Brain Sci. 36, 255–274.

Purushothaman, G., Karayiannis, N., 1997. Quantum neural networks (QNNs): inherently fuzzy feedforward neural networks. IEEE Trans. Neural Netw. 8(3), 679–693.

Raina, R., Madhavan, A., Ng, A., 2009. Large-scale deep unsupervised learning using graphics processors. In: Proceedings of ICML-09, 26th Annual International Conference on Machine Learning.

Rätsch, G., Onoda, T., Müller, K.-R., 2001. Soft margins for AdaBoost. Mach. Learn. 42(3), 287–320.

Rebentrost, P., Mohseni, M., Lloyd, S., 2013. Quantum support vector machine for big feature and big data classification. arXiv:1307.0471.

Roland, J. Cerf, N.J., 2002. Quantum search by local adiabatic evolution. Phys. Rev. A 65, 042308.

Rønnow, T.F., Wang, Z., Job, J., Boixo, S., Isakov, S.V., Wecker, D., Martinis, J.M., Lidar, D.A., Troyer, M., 2014. Defining and detecting quantum speedup. arXiv:1401.2910.

Rosenblatt, F., 1958. The perceptron: a probabilistic model for information storage and organization in the brain. Psychol. Rev. 65(6), 386–408.

Rumelhart, D., Hinton, G., Williams, R., 1986. Learning Internal Representations by Error Propagation. MIT Press, Cambridge, MA.

Rumelhart, D., Widrow, B., Lehr, M., 1994. The basic ideas in neural networks. Commun. ACM 37(3), 87–92.

Sasaki, M., Carlini, A., Jozsa, R., 2001. Quantum template matching. Phys. Rev. A 64(2), 022317.

Sasaki, M., Carlini, A., 2002. Quantum learning and universal quantum matching machine. Phys. Rev. A 66, 022303.

Sato, I., Kurihara, K., Tanaka, S., Nakagawa, H., Miyashita, S., 2009. Quantum annealing for variational Bayes inference. In: Proceedings of UAI-09, 25th Conference on Uncertainty in Artificial Intelligence, pp. 479–486.

Scarani, V., 2006. Feats, features and failures of the PR-box. AIP Conf. Proc. 884, 309–320.

Schaller, G., Mostame, S., Schützhold, R., 2006. General error estimate for adiabatic quantum computing. Phys. Rev. A 73, 062307.

Schapire, R.E., 1990. The strength of weak learnability. Mach. Learn. 5(2), 197–227.

Schölkopf, B., Smola, A.J., 2001. Learning with Kernels: Support Vector Machines, Regularization, Optimization, and Beyond. MIT Press, Cambridge, MA.

Sebastiani, F., 2002. Machine learning in automated text categorization. ACM Comput. Surv. 34(1), 1–47.

Sentís, G., Calsamiglia, J., Muñoz Tapia, R., Bagan, E., 2012. Quantum learning without quantum memory. Sci. Rep., 2, 1–8.

Servedio, R.A., Gortler, S.J., 2001. Quantum versus classical learnability. In: Proceedings of CCC-01, 16th Annual IEEE Conference on Computational Complexity, pp. 138–148.

Servedio, R.A., Gortler, S.J., 2004. Equivalences and separations between quantum and classical learnability. SIAM J. Comput. 33(5), 1067–1092.

Settles, B., 2009. Active learning literature survey. Technical Report 1648, University of Wisconsin, Madison.

Shalev-Shwartz, S., Shamir, O., Sridharan, K., 2010. Learning kernel-based halfspaces with the zero-one loss. In: Proceedings of COLT-10, 23rd Annual Conference on Learning Theory, pp. 441–450.

Shawe-Taylor, J., Cristianini, N., 2004. Kernel Methods for Pattern Analysis. Cambridge University Press, Cambridge.

Shin, S.W., Smith, G., Smolin, J.A., Vazirani, U., 2014. How "quantum" is the D-wave machine? arXiv:1401.7087.

Shor, P., 1997. Polynomial-time algorithms for prime factorization and discrete logarithms on a quantum computer. SIAM J. Comput. 26, 1484.

Silva, J., Marques, J., Lemos, J., 2006. Selecting landmark points for sparse manifold learning. Adv. Neural Inf. Process. Syst. 18, 1241–1247.

Smola, A., Schölkopf, B., Müller, K., 1998. The connection between regularization operators and support vector kernels. Neural Netw. 11(4), 637–649.

Sörensen, K., 2013. Metaheuristics—the metaphor exposed. International Transactions in Operational Research. http://dx.doi.org/10.1111/itor.12001.

Steinbach, M., Karypis, G., Kumar, V., 2000. A comparison of document clustering techniques. In: KDD Workshop on Text Mining.

Steinwart, I., 2003. Sparseness of support vector machines. J. Mach. Learn. Res. 4, 1071–1105.

Stempfel, G., Ralaivola, L., 2009. Learning SVMs from sloppily labeled data. In: Alippi, C., Polycarpou, M., Panayiotou, C., Ellinas, G. (Eds.), Proceedings of ICANN-09, 19th International Conference on Artificial Neural Networks, pp. 884–893.

Sun, J., Feng, B., Xu, W., 2004. Particle swarm optimization with particles having quantum behavior. In: Proceedings of CEC-04, Congress on Evolutionary Computation, vol. 1, pp. 325–331.

Suykens, J.A., Vandewalle, J., 1999. Least squares support vector machine classifiers. Neural Process. Lett. 9(3), 293–300.

Tenenbaum, J., Silva, V., Langford, J., 2000. A global geometric framework for nonlinear dimensionality reduction. Science 290(5500), 2319–2323.

Trugenberger, C.A., 2001. Probabilistic quantum memories. Phys. Rev. Lett. 87, 067901.

Trugenberger, C.A., 2002. Phase transitions in quantum pattern recognition. Phys. Rev. Lett. 89, 277903.

Valiant, L.G., 1984. A theory of the learnable. Communn. ACM 27(11), 1134–1142.

Van Dam, W., Mosca, M., Vazirani, U., 2001. How powerful is adiabatic quantum computation? In: Proceedings of FOCS-01, 42nd IEEE Symposium on Foundations of Computer Science, pp. 279–287.

Vapnik, V.N., Chervonenkis, A.Y., 1971. On the uniform convergence of relative frequencies of events to their probabilities. Theor. Probab. Appl. 16(2), 264–280.

Vapnik, V., 1995. The Nature of Statistical Learning Theory. Springer, New York, NY.

Vapnik, V., Golowich, S., Smola, A., 1997. Support vector method for function approximation, regression estimation, and signal processing. Adv. Neural Inf. Process. Syst. 9, 281.

Ventura, D., Martinez, T., 2000. Quantum associative memory. Inform. Sci. 124(1), 273–296.

Vidick, T., Wehner, S., 2011. More nonlocality with less entanglement. Phys. Rev. A 83(5), 052310.

Weinberger, K., Sha, F., Saul, L., 2004. Learning a kernel matrix for nonlinear dimensionality reduction. In: Proceedings of ICML-04, 21st International Conference on Machine learning, pp. 106–113.

Weinstein, M., Horn, D., 2009. Dynamic quantum clustering: a method for visual exploration of structures in data. Phys. Rev. E 80(6), 066117.

Weston, J., Mukherjee, S., Chapelle, O., Pontil, M., Poggio, T., Vapnik, V., 2000. Feature selection for SVMs. Adv. Neural Inf. Process. Syst. 13, 668–674.

Wiebe, N., Berry, D., Høyer, P., Sanders, B.C., 2010. Higher order decompositions of ordered operator exponentials. J. Phys. A Math. Theor. 43(6), 065203.

Wiebe, N., Kapoor, A., Svore, K.M., 2014. Quantum nearest neighbor algorithms for machine learning. arXiv:1401.2142.

Wittek, P., Tan, C.L., 2011. Compactly supported basis functions as support vector kernels for classification. IEEE Trans. Pattern Anal. Mach. Intell. 33(10), 2039–2050.

Wittek, P., 2013. High-performance dynamic quantum clustering on graphics processors. J. Comput. Phys. 233, 262–271.

Wolpert, D.H., 1992. Stacked generalization. Neural Netw. 5(2), 241–259.

Wolpert, D.H., Macready, W.G., 1997. No free lunch theorems for optimization. IEEE Trans. Evol. Comput. 1(1), 67–82.

Yang, Y., Chute, C., 1994. An example-based mapping method for text categorization and retrieval. ACM Trans. Inf. Syst. 12(3), 252–277.

Yang, Y., Liu, X., 1999. A re-examination of text categorization methods. In: Proceedings of SIGIR-99, 22nd International Conference on Research and Development in Information Retrieval, pp. 42–49.

Young, A.P., Knysh, S., Smelyanskiy, V.N., 2010. First-order phase transition in the quantum adiabatic algorithm. Phys. Rev. Lett. 104, 020502.

Yu, H., Yang, J., Han, J., 2003. Classifying large data sets using SVMs with hierarchical clusters. In: Proceedings of SIGKDD-03, 9th International Conference on Knowledge Discovery and Data Mining, pp. 306–315.

Yu, Y., Qian, F., Liu, H., 2010. Quantum clustering-based weighted linear programming support vector regression for multivariable nonlinear problem. Soft Comput. 14(9), 921–929.

Zak, M., Williams, C.P., 1998. Quantum neural nets. Int. J. Theor. Phys. 37(2), 651–684.

Zaki, M.J., Meira Jr., W., 2013. Data Mining and Analysis: Fundamental Concepts and Algorithms. Cambridge University Press, Cambridge.

Zhang, L., Zhou, W., Jiao, L., 2004. Wavelet support vector machine. IEEE Trans. Syst. Man Cybern. B Cybern. 34(1), 34–39.

Zhang, T., Yu, B., 2005. Boosting with early stopping: convergence and consistency. Ann. Stat. 33(4), 1538–1579.

Zhou, R., Ding, Q., 2008. Quantum pattern recognition with probability of 100%. Int. J. Theor. Phys. 47, 1278–1285.

Printed in the United States
By Bookmasters